More Than Human

EMBRACING THE PROMISE OF BIOLOGICAL ENHANCEMENT

Ramez Naam

First edition published 2005 by Broadway Books, a division of Random House.

ISBN: 978-0-557-58233-4

Table of Contents

Introduction - Healing and Enhancing

Southern Illinois University Medical School rises out of the cornfields that surround the state capital of Springfield, three hours south of Chicago. Prairie dogs, field mice, and other animals scurry around the rolling acres of the countryside. In a lab in a nondescript building on the campus, a thin, balding, smiling man named Andrzej Bartke has been investigating the causes of human aging. His goal is to learn to delay the onset of heart disease, cancer, frailty, and other problems that come with aging - to increase the human health span, as he calls it.

Bartke holds two mice in his cupped hands. One, a close cousin of the mice that scamper outside, is of a breed that live for two and a half years on average before succumbing to told age. The other mouse has been genetically engineered to live longer. The oldest of its kind died just a week short of its fifth birthday, twice as long as a genetically normal mouse, as if a human had lived to the age of 200. The difference between the breeds is just a change to a single gene, a gene that exists in everything from yeast to mice to men and women. In seeking to cure disease, Bartke and his colleagues have opened the possibility for something more – a lengthening of the human lifespan to superhuman levels.

The same basic story has repeated itself dozens of times in the past decade. Researchers looking for ways to heal the sick or injured or preserve the health of the elderly have stumbled on techniques that might enhance human abilities. At Princeton University researchers looking for ways to stave off Alzheimer's disease have created mice who can learn things more than twice as fast as their genetically normal peers. Their research is pointing to techniques that could improve human learning as well.

In a lab at the University of Pennsylvania, researchers trying to find a cure for muscular dystrophy have produced mice that are super strong. They can climb things and lift things that their normal siblings can't. At some point in their lives, each has been injected with extra copies of a gene for building muscle. The same gene plays a role in human muscle building.

At Duke University, scientists looking for ways to help amputees and paralytics have implanted a group of monkeys with electrodes in their brains. The monkeys can move mechanical arms *just by thinking about it*, as if those robots were parts of their bodies.

In Lisbon, Portugal, there's a group of blind men and women who can now see. In place of eyeglasses they wear cameras connected to electrodes implanted in the visual parts of their brains. Some of them were blind for twenty years or more before the surgery. The same research that gives them sight could beam images from one person's mind into another.

The scientists conducting these studies have a common goal - to heal sick and injured men and women. Along the way, they're discovering a common fact. To heal our minds and bodies we must understand them. And in understanding we gain the power to improve. That power could give us better learning and memory, better muscles, longer lives, and much more. By unraveling how our minds and bodies function, biotechnology could give us the power to sculpt any aspect of ourselves - how we think, how we feel, how we look, how we communicate with one another.

Not everyone is convinced that this power to alter ourselves is a benefit to society. There are those who oppose too much tinkering with the human mind and body. These concerns have been voiced across the political spectrum. In 2000, President George W. Bush created the President's Council on Bioethics to advise him on issues of biotechnology. To head the council, he appointed Leon Kass, a conservative University of Chicago professor with a twenty-five year history of opposing infertility treatments, cosmetic surgery, organ transplantation, and other technologies that, in his view, violate the natural order of things. Under Kass's direction, the council has released report after report condemning the use of new bioetechnologies to alter the human mind or body.

The council's 2004 report, *Beyond Therapy*, argues that genetic and reproductive technologies undermine the value of life and disrupt the natural relationship between parents and children; That slowing human aging would stagnate society as the old cling stubbornly to power; That instant performance enhancers would reduce human drive and hard work; That technologies for sculpting the human mind would threaten our sense of identity; That enhancement techniques might widen the gap between rich and poor; That the safety of these techniques can not be assured; and that they could be used by the powerful to coerce or control the weak.

Ultimately the council argues that we ought to revere our given, natural state, that to seek to improve on what we have is hubris, and that biotechnological alteration of our minds and bodies threatens our unique human dignity.

In his book *Our Posthuman Future*, bio-ethics council member Francis Fukuyama echoes this sentiment. "There are good prudential reasons to defer to the natural order of things and not to think that human beings can easily improve on it."

Given these concerns, how should society react? Fukuyama believes that "There are certain things that should be banned outright. One of them is reproductive cloning... If we get used to cloning in the near term, it will be much harder to oppose germ-line [genetic] engineering for enhancement purposes in the future."

In fact, Fukuyama would like to restrict more than just genetic technologies. Elsewhere in his book he argues that the governments need to "draw red lines" around technologies in general, "to distinguish between therapy and enhancement, directing research toward the former while putting restrictions on the latter."

Along those lines liberal philosopher and bioethicists George Annas has called genetic engineering a "crime against humanity" and argued for a UN treaty prohibiting it. William Kristol, publisher and editor of the conservative *Weekly Standard*, sees things similarly. "A ban on cloning would only be a first step down the road of responsibility and self-government – but it would be an important step."

Environmentalist Bill McKibben goes farther, calling for a halt on even the basic scientific research that might enhance human abilities. In his book *Enough*, he writes "We need to do an unlikely thing: we need to survey the world we now inhabit and proclaim it good. Good enough. [....] Enough intelligence. Enough capability. Enough."

In short, a chorus of voices now argue that we should strive to preserve the status quo, that we should opt for stability over change, for the known over the unknown. To do this, as Fukuyama says, "We should use the power of the state" to restrict access to technologies that might undermine our current notions of humanity, that might allow individuals to surpass the mental and physical limitations we now know.

This book is based on a very different premise – that rather than fearing change, we ought to embrace it, that rather than prohibiting these technologies, society ought to focus on spreading the power to alter our own minds and bodies to as many people as possible; that rather than imposing a rigid view of what it means to be human on humanity, we ought to trust billions of individuals and families to make that decision for themselves. In making that case, I'm going to show you four things.

FIRST, THERE ARE good pragmatic reasons to embrace human enhancement. Enhancement itself is a fuzzy concept. Scientifically there's no clear line between healing and enhancing. To ban the development of research that could keep us young, that could improve our memories, that could wire our minds together, or that could enhance us in other ways, we would have to ban the most promising research into curing Alzheimer's disease, or reducing heart disease and cancer, or restoring sight to the blind and motion to the paralyzed. If we are to ban such improvements to our quality of life, we had better have strong evidence that the research poses a greater threat to society than the benefits it brings.

Those benefits – even beyond curing disease – are concrete and measurable. Slowing aging – keeping people younger longer – would slow the rise in worldwide health spending, and help stave off the demographic crunch of an aging population around the world. Improving human memory, attention, and communication abilities would increase productivity. It would lead to new scientific discoveries and faster innovation, economic growth and new technologies we can't anticipate today.

SECOND, A BAN wouldn't solve any problems – it would only make matters worse. As Steven Hyman, former director of the National Institute of Mental Health, told the President's Council on Bio-Ethics that "Regulation that flies in the face of reality ... decreases respect for regulation and fails," and that, "if we can find medications which will enhance our performance, lengthen our life, decrease the stress ... no regulation in the world would keep them from general use."

Prohibition wouldn't stop people from seeking enhancement any more than the War on Drugs has stopped people from seeking recreational chemicals. A ban would only make enhancement more dangerous and more expensive by creating a black market with an artificially high price, further separating the rich and poor. It would add risk to those who sought enhancement, by removing any possibility of safety regulations. Furthermore a ban would hinder scientific progress as long term studies of the effects of these technologies would become impossible. It would harm society as biotech changes to our minds and bodies moved underground, where the public would have less visibility into the research going on.

THIRD, THE DEBATE over human enhancement is at heart a debate over human freedom. Should individuals and families have the right to alter their own minds and bodies? Or should that power be held by the state?

In a democratic society, it's every man and woman who should determine such things, and not the state. Western democracies are founded on the principle that governments exist to protect the freedoms of individuals, or so the Declaration of Independence claims:

> "We hold these truths to be self-evident, that all men are created equal, that they are endowed by their Creator with certain unalienable Rights,

> that among these are Life, Liberty and the pursuit of Happiness. --That to secure these rights, Governments are instituted among Men"

Yet beneath the arguments of bio-conservatives is the assumption that a small cadre of elites – legislators, regulators, and professional ethicists – know what's best, that they can make better decisions on behalf of millions or billions of people than those individuals could make for themselves.

In an essay on the perils of lengthening human life, for example, Bush's top ethicist Leon Kass writes, "The finitude of life is a blessing for every individual, whether he knows it or not." Is this really for Kass to decide?

Governments are instituted to *secure* individual rights, not to restrict them. This is both an issue of principle and of good governance: The last century of history suggests that nations that embrace the freedoms of individuals thrive, while nations that limit individual rights fall behind and fail, often in monstrous ways.

In 1992, before the publication of his anti-biotech treatise *Our Posthuman Future,* Francis Fukuyama wrote a book titled *The End of History*. Fukuyama argued, as other historians have, that the twentieth century saw the victory of liberal democracy over other forms of government, and specifically totalitarianism.

At the start of the twentieth century, systems like communism were formed around the belief that government could make better decisions for the people than the people could for themselves. Democracies were formed around more or less the opposite principle – that individuals making decisions for themselves would produce, on the whole, better results than central control.

At the end of the 20th century, Fukuyama wrote, it was clear that liberal democracy had won. There could no longer be any doubt that free societies, societies where individuals made most of the decisions for themselves, could economically, scientifically, and technologically outcompete centrally controlled societies. Just as importantly, there could no longer be any doubt that individuals living in free societies were happier and better off than anywhere else in the world.

The triumph of democracy has its roots in two simple phenomena. Individual people in general want to improve on their situation. They want to live longer, remain in good health, increase their capabilities, increase their options, increase the well being of their children, and so on. Given the choice they'll work in this direction.

In addition, millions of individuals weighing costs and benefits simply have greater collective intelligence, better collective judgment, then a small number of central regulators and controllers. People aren't perfect decision makes by any means. Individuals make a tremendous number of mistakes, but on the whole the distributed decision making power of the masses is more effective at moving society towards greater welfare than the decisions made by a small elite.

Those two simple facts explain why a market economy driven by the often rather shallow desires of consumers nonetheless out-competes and out-innovates a more centrally controlled economy. They explain why political systems based on votes cast by a public with widely varying views are nonetheless more stable and more effective at preserving the rights of citizens than political systems based on monarchy or dictatorship.

If the lessons of the 20th century hold true into the 21st, then the nations which attempt to control the minds and bodies of their people will fall behind, while nations that embrace human freedom to alter our own minds and bodies will thrive.

FINALLY, FAR FROM being unnatural, the drive to alter and improve on ourselves is a fundamental part of who we are. As a species we've always looked for ways to be faster, stronger, smarter, and longer lived. Every past enhancement - from transfusions to vaccinations to birth control - has been called unnatural or immoral. Yet over time we've become accustomed to these new levels of control over our minds and bodies, and used them for the betterment of ourselves, our families, and our world.

New discoveries have at times caused upheavals in our notion of who we are and what our lives mean. Galileo's discovery that the Earth is not the center of the universe challenged our 17th century conception of our place in the cosmos. Darwin's theory of evolution challenged our 19th century conception of man's place in nature. Now science is challenging the idea that our mental and physical forms are fixed. We can, if we choose to, change them.

Yet our lives today are no less meaningful and no less challenging than they were centuries ago. If anything, greater knowledge and power over ourselves and our world has enriched us. Technologies that have increased our abilities - telephones, cars, books, and medicine - have increased our opportunities to connect with one another, to learn about our world, to contribute to the rest of humanity, and thus to lead more meaningful lives.

The only reason we live lives of such great comfort and potential today is that throughout history there have been men and women who've refused to accept the natural order of things. From the mastery of fire, to agriculture, to written language, to eye glasses, to antibiotics, previous generations constantly improved their world, their minds, and their bodies in search of a better life for themselves and their families. That constant quest to improve their situation has passed its benefits down to us. The men and women striving for better lives probably never imagined the impact their efforts would have decades, centuries, or millennia down the line. Whatever efforts we make to improve our own lives and the lives of our children will benefit future generations in ways that we ourselves can't anticipate.

In the end, this search for ways to enhance ourselves is a natural part of being human. It's been a force in history as far back as we can see. It's been selected for by millions of years of evolution. It's wired deep in our genes - a natural outgrowth of our human intelligence, curiosity, and drive. To turn our back on this power would be to turn our back on our true natures. Embracing our quest to understand and improve on ourselves doesn't call into question our humanity - it reaffirms it.

Chapter 1 – Designer Bodies

In 1989, Raj and Van DeSilva were desperate. At the age of four, their daughter Ashanti was dying. She was born with a crippled immune system, a consequence of a deadly gene she carried.

Every human being has around thirty thousand genes. In fact, we have two copies of each of those genes - one inherited from our mother, the other from our father. Our genes tell our cells what proteins to make, and when.

Each protein is a tiny molecular machine. Every cell in your body is built out of millions of these little machines, working together in precise ways. Proteins break down food, ferry energy to the right places, and form scaffoldings that keep our cells in the right shape. Proteins synthesize messenger molecules to pass signals in the brain and other proteins form receptors to receive those signals. Even your ribosomes - the machines inside each of your cells which build new proteins -are themselves made up of other proteins.

Ashanti DeSilva inherited two broken copies of the gene for a protein called adenoside deaminase (ADA). If she had had one broken copy, she would have been fine. The other copy of the gene would have made up the difference. With two broken copies, her body didn't have the right instructions to manufacture ADA at all.

ADA is a crucial part of our resistance to disease. Without it, special white blood cells called T cells die off. Without T cells, ADA-deficient children are wide open to viruses and bacteria. They have what's called Severe Combined Immune Deficiency disorder (SCID), more commonly known as "Bubble Boy" disease.

With a weak immune system, the outside world is threatening. Every person you touch, share a glass with, or share the same air with is a potential source of dangerous pathogens. Without the ability to defend herself, Ashanti was largely confined to her home.

The standard treatment for ADA-deficiency is frequent injections of PEG-ADA, a synthetic form of the ADA enzyme. PEG-ADA can mean the difference between life and death for an ADA-deficient child. Unfortunately, while it usually produces a rapid improvement when first used, children tend to respond less and less to the drug each time. Ashanti DeSilva started receiving PEG-ADA injections at the age of two. Initially she responded well. Her T-cell count rose sharply and she developed some resistance to disease. But by the age of four, she was slipping away, no longer responding strongly to her injections. If she was to live, she'd need something more than PEG-ADA. The only other option at the time, a bone marrow transplant, was ruled out by the lack of matching donors.

In early 1990, while Ashanti's parents were searching frantically for help, geneticist French Anderson at the National Institutes of Health was seeking permission to perform the first gene therapy trials on humans. Anderson, an intense fifth degree blackbelt in tae-kwon doe and respected researcher in the field of genetics, wanted to show that he could treat genetic diseases caused by faulty copies of genes by inserting new, working copies of the same gene.

Scientists had already shown that it was possible to insert new genes into plants and animals. Genetic engineering got its start in 1972, when Stanley Cohen and Herbert Boyer first met at a scientific conference in Hawaii. The conference was about plasmids - small circular loops of DNA in which bacteria carry their genes. Cohen, then a professor at Stanford, had been working on ways to insert new plasmids into bacteria. Boyer's lab at the University of California in San Francisco had recently discovered restriction enzymes - molecular tools that could be used to slice and dice DNA at specific points.

Over a dinner of hot pastrami and corned beef sandwiches, the two researchers concluded that their technologies complemented one another. Boyer's restriction enzymes could isolate specific genes which Cohen's techniques could then deliver to bacteria. Combined, the two discoveries enabled the researchers to alter the genes of the bacteria. In 1973, just four months after meeting, Cohen and Boyer inserted a new gene into the *e. coli* bacterium.

For the first time, humans were tinkering directly with the genes of another species. The field of genetic engineering was born. Boyer would go on to found Genentech, the world's first biotech company. Cohen would go on to do other critical work in biology.

Building on Cohen and Boyer's work with bacteria, hundreds of scientists went on to find ways to insert new genes into plants and animals. The hard work of genetically engineering these higher organisms is getting the new gene into the cells. To do this one needs a gene vector - a way to get the gene to the right place. Most researchers use gene vectors provided by nature - viruses. In some ways, viruses are an ideal tool for ferrying genes into a cell, because that's already one of their main functions. Viruses are cellular parasites. Unlike plant or animal cells, or even bacteria, viruses can't reproduce themselves. Instead, they penetrate into cells and implant their viral genes. Those genes instruct the cell to make more of the virus, one protein at a time.

Early genetic engineers realized that they could use viruses to deliver whatever genes they wanted. Instead of delivering the genes to create more virus, they could be modified to deliver a gene chosen by a scientist. Modified viruses were pressed into service as genetic "trucks", carrying a payload of genes loaded onto them by researchers. Not only do these viruses deliver the desired genes, they also don't spread from cell to cell, because they don't carry the genes necessary for the cell to make new copies of the virus.

By the late 1980s researchers had used this technique to alter the genes of dozens of species of plants and animals - tobacco plants that glow, tomatoes that could survive freezing, corn resistant to pesticides. French Anderson and his colleagues reasoned that one could do the same in a human being. Given a patient who lacked a gene crucial to health, one ought to be able to give that person copies of the missing gene. This is what Anderson proposed to do for Ashanti.

Starting in June of 1988, Anderson's proposed clinical protocols went through tremendous scrutiny and endured more than a little hostility. His first protocol was reviewed by both the National Institutes of Health (NIH) and the Food and Drug Administration (FDA). Over a period of seven months, seven regulatory committees conducted fifteen meetings and twenty hours of public hearings to assess the proposal.

In early 1990, Anderson and his collaborators received the final approval from the NIH's Recombinant DNA Advisory Committee and had cleared all legal hurdles. By spring, they had identified Ashanti as a potential patient. Would her parents consent to an experimental treatment? Of course there were risks to the therapy. Yet without it, Ashanti would face a life of seclusion and probable death in the next few years. Given these odds, her parents opted to try the cure. As Raj DeSilva told the *Houston Chronicle*, "What choice did we have?"

Ashanti and her parents flew to the NIH Clinical Center at Bethesda, Maryland. There, over the course of twelve days, Anderson and his colleagues Michael Blaese and Kenneth Culver slowly extracted some of Ashanti's blood cells. Safely outside the body, the cells would have a new, working copy of the ADA gene inserted into them by a hollowed out virus. Finally, starting on the afternoon of September 14, Culver injected the cells back into Ashanti's body.

The gene therapy had roughly the same goal as a bone marrow transplant - give Ashanti a supply of her own cells that could produce the ADA enzyme. Unlike a bone marrow transplant, gene therapy had no risk of rejection. The cells Culver injected back into Ashanti's bloodstream were her own, and her body recognized them as such.

The impact of the gene therapy on Ashanti was striking. Within six months her T cell count had risen to normal levels. Over the next two years her health continued to improve, allowing her to enroll in school, venture out of the house, and lead a fairly normal childhood.

No one will say that Ashanti is completely cured. She still takes a low dose of PEG-ADA. While normally the size of doses would grow with age, her doses have remained fixed at her four-year-old level. It's possible that she could be taken off the PEG-ADA therapy entirely, but her doctors don't think it's yet worth the risk. The fact that she's alive today – let alone healthy and active – is due to her gene therapy. And in going through that therapy, she helped prove a crucial point - genes can be inserted into humans to cure genetic diseases.

From Healing To Enhancing

Following Ashanti's treatment, the field of gene therapy blossomed. Since 1990 hundreds of labs have begun experimenting with gene therapy as a technique to cure disease, and more than five hundred human trials involving over four thousand patients have been launched. Researchers have shown that it may be possible to use gene therapy to cure diabetes, sickle cell anemia, several kinds of cancer, Huntington's disease, and even to open blocked arteries.

While the goal of gene therapy researchers is to cure disease, gene therapy could also be used to boost human athletic performance. In many cases, the same research that is focused on saving lives has also shown that it can enhance the abilities of animals, with the suggestion that it could enhance men and women as well.

Consider the user of gene therapy to combat anemia. Circulating through your veins are trillions of red blood cells. Pumped by your heart, they serve as the delivery system of your body, ferrying oxygen from the lungs to rest of your tissues, and carrying carbon dioxide back out to the lungs. Without enough red blood cells, you can't function. Your muscles can't get enough oxygen to produce force, and your brain can't get enough oxygen to think clearly. Hundreds of thousands of people worldwide live with this anemia, and with the lethargy and weakness that it brings. In the US at least eighty five thousand patients are severely anemic due to kidney failure. Another fifty thousand AIDS patients are anemic due to side effects of the HIV drug AZT.

In 1985, researchers at a Thousand Oaks California based biotech company, Amgen, started looking for a way to treat anemia isolated the gene for EPO. EPO is a hormone found in the human body. Your kidneys produce

EPO in response to low oxygen levels. EPO, in turn, causes your body to generate more red blood cells. For a patient whose kidneys have failed, injections of Amgen's synthetic EPO can take up some of the slack. The drug is a lifesaver, so popular that the worldwide market for it is as high as $5 billion per year, and therein lies the problem. Three injections of EPO a week adds up. Patients who need this kind of therapy end up paying $7,000 to $9,000 a year. In poor countries struggling to even pay for HIV drugs like AZT, the added burden of paying for EPO to offset the side effects just isn't feasible.

What if there was another way? What if the body could be instructed to produce more EPO on its own, to make up for that lost to kidney failure or AZT? That's the question Jeffrey Leiden asked himself in the mid 1990's. In 1997, the University of Chicago professor and his colleagues performed the first animal study of EPO gene therapy. They injected lab monkeys and mice with a virus carrying an extra copy of the EPO gene. The virus, in turn, infected some tiny proportion of the cells in the mice and monkey. The cells began to produce extra EPO, causing the animals' bodies to create more red blood cells. In principle, this was no different from injecting extra copies of the ADA gene into Ashanti, except in this case the animals already had two working copies of the EPO gene. The one being inserted into some of their cells was a third copy, and if it worked it would boost their levels of EPO production beyond the norm for their species.

That's just what happened. After just a single injection, the animals began producing more EPO, and their red blood cell counts soared. The mice went from a hematocrit of 49 percent (meaning that 49 percent of their blood by volume was red blood cells), to 81 percent. The monkeys went from 40 percent to 70 percent. At least two other biotech companies, Chiron and Ariad, have produced similar results in baboons and monkeys respectively.

The increase in red blood cell count is impressive, but doctors can already produce that in patients with injections of EPO itself. They don't need to go to the added trouble of using a virus to deliver the EPO gene to the patient's body. The real advantage of gene therapy is in duration. EPO injections have to be repeated three times a week. EPO gene therapy on the other hand, could be administered as rarely as every few months, or even just once for the patient's entire lifetime.

The research bears this out. In Leiden's original experiment, the mice (who each received just one shot) showed higher red blood cell counts for a year. In the monkeys the effects lasted for twelve weeks. Chiron's baboons enjoyed the benefits of their single injection for the entire duration of the seven month experiment. The monkeys in the Ariad trial, who went through gene therapy more than four years ago, still show higher red blood cell counts today.

This is a key difference between drug therapy and gene therapy. Drugs sent into the body have their effect for a while, and then are eventually broken up or passed out. Gene therapy, on the other hand, gives the body the ability to manufacture the needed protein or enzyme or other bodily chemical itself. The new genes can last for a few weeks or can become a permanent part of the patient's genome.

The difference in duration is a result of the kind of gene vector used and where it delivers its payload of DNA. Almost all of the DNA you carry is on twenty three pairs of chromosomes that are inside the nucleus of your cells. The nucleus forms a protective barrier shielding your chromosomes from damage. It also contains sophisticated DNA repair mechanisms that patch up most of the damage that does occur.

Insertional gene vectors penetrate all the way into the nucleus of the cell, and splice the genes they carry into the chromosomes. From that point on, the new genes get all the benefits your other genes enjoy. The new genes are shielded from most of the damage that can happen inside your cells. If the cell divides, the new genes get copied to the daughter cells, just like the rest of your DNA. Insertional vectors make more or less permanent changes to your genome.

Non-insertional vectors, on the other hand, don't make it into the nucleus of your cells. They don't splice the new genes they carry into your chromosomes. Instead, they deliver their payload of DNA and leave it floating around inside your cells. The new DNA still gets read by the cell. It still instructs the cell to make new proteins. But it doesn't get copied when the cell divides. And over time it suffers from wear and tear, until eventually it breaks up, and its effects end.

The difference in durations gives us choices. We can choose to make a temporary change with a drug, which will wear off in a few hours or days. We can make a semi-permanent change with non-insertional gene therapy, and the effects will last for weeks or months depending on the genes and the type of cell infected. Or we can choose make a permanent change by inserting new genes directly into your genome.

We'll come back to talk about why each of these three options are appropriate at certain times. For now, in the context of EPO, the idea of semi-permanent or permanent change has definite advantages. It cuts down on the need for frequent injections, which means that the gene therapy approach can end up much cheaper than the drug therapy approach.

Boosting Athletic Performance

If you've heard of EPO, it may not have been for the drug's medical uses. The Tour de France is the most famous and prestigious cycling race in the world. In 1998 while the race was ongoing, customs officials at the border between Belgium and France found more than 400 doses of EPO the car of Willy Voet, the massage therapist for Team Festina – one of the prime contenders in the race. Within days the mostly Swiss and French team were disqualified.

Injections of EPO cause your body to produce more red blood cells, whether you're anemic or not. With more red blood cells, your body can carry more oxygen. With greater oxygen-carrying capacity, your endurance increases. If you're an elite athlete, injections of synthetic EPO can boost your performance in distance running, swimming, and bicycling events by as much as 10 percent. In these sports, where victory is often snatched by a few fractions of a percent, that's an incredible advantage.

EPO isn't the first drug athletes have used to boost performance. In the 1988 Olympics, sprinter Ben Johnson had his gold medal stripped from him after testing positive for anabolic steroids. Baseball player and 1999 home-run record setter Mark McGuire has admitted to using the steroid androstenedione. Other athletes have used human growth hormone, methamphetamines, transfusions of additional human blood or hemoglobin, or synthetic blood additives called PFCs.

Some of these athletic performance boosters have been adopted by amateur athletes. Recreational weight lifters, for example, have been known to use steroids and human growth hormone. Yet the most effective techniques have substantial health risks that discourage widespread use. Anabolic steroids can cause heart disease, high blood pressure, kidney problems, reduced sperm count and impotence in men, increased body hair and other masculine features in women, and increased aggression in both. EPO is generally safe at low doses, but too much can thicken the blood to the point where the heart can no longer pump, leading to heart attacks and death.

The inconvenience of these drugs also discourages widespread use. Like EPO, many of the other performance enhancing drugs require multiple injections per week. Frequent doses mean that a user has to keep a supply of a usually illegal or prescription-only substance on hand. Most people looking for a small boost to their tennis game aren't willing to subject themselves to frequent needle-pricks or the risks of regularly buying and possessing a controlled substance.

Gene therapy changes the rules. EPO gene therapy would make it easier for athletes to cheat by reducing the risk of detection. Synthetic EPO is already hard to detect, because it's almost identical to the EPO produced by the body. Athletes using EPO aren't caught by blood tests. They're only discovered if they're caught in possession of EPO, or with EPO in the possession of their trainers. If an athlete could take a shot months before the event, instead of at the event itself, the risk of being caught in possession of the drug would drop dramatically.

The bigger impact of EPO gene therapy may be on recreational athletes. Few weekend hikers, bikers, or soccer players care enough about their performance to inject themselves three times a week. But what about an injection every few months? Or just one injection in a lifetime?

With integrating gene vectors, it looks like it will be possible to give a person just a single injection and have that injection increase their red blood cell count, their aerobic fitness, their endurance, *for life*.

Of course, EPO gene therapy would come with risks. First there are the risks of gene therapy itself - the possibility that the newly inserted gene could disrupt the function of an existing gene, or that the body's immune system might respond. Yet as we discussed in the last chapter, there's a tremendous medical incentive to improve the safety and effectiveness of gene therapy, and a number of promising techniques have already shown success in human cells and live animals.

The more serious risk of adding an EPO gene is the risk of EPO itself. EPO has already been implicated in the deaths of quite a few athletes, including around twenty Dutch cyclists in the early 1990s. These athletes died because they took too much EPO. They pushed their red blood cell count to the extreme. They thickened their blood to the point where their heart could no longer pump it. Since thenthe number of EPO deaths has declined as athletes have learned more about the drug. Aware of the risks, they can carefully moderate their own use of the drug, cutting back a bit if their blood becomes too thick.

Biotech firms are also aware of these risks. They've come up with ingenious ways to throttle the amount of EPO the genes added through gene therapy produce. These techniques relate to the EPO gene's *promoter*. The promoter region of the gene is the part that tells the cell when it should actually create some of the protein the gene describes.

Some gene therapy trials have the goal of producing as much of the protein the inserted gene describes as possible. In these cases, researchers use promoters that are simply "on" all the time. With EPO this isn't an option, since too much of the EPO protein can kill.

Instead of using an "always-on" promoter, researchers have come up with two different systems to control the activity of the gene. Researcher Katie Binley and her colleagues at Oxford Biomedica in the United Kingdom have developed a promoter tied to low oxygen levels in the blood. If EPO levels are high, blood oxygen will remain high, and the gene won't be activated to produce more EPO. Another company, Ariad Gene Therapies, has created a promoter that's triggered by a modified version of the drug rapamycin. The modified rapamycin can be taken in the form of a pill - more convenient than an injection. Once it's taken, it activates the activity of the extra EPO genes for around a month. That makes the gene therapy plus pill approach quite a bit more convenient than thrice weekly injections, while still providing some control over red blood cell count.

Ariad's promoter is part of its gene therapy technique, which will soon enter human trials. By adding a degree of control over the EPO gene, these gene promoters should increase the safety of EPO gene therapy. They also illustrate some of the flexibility of gene therapy. We can control how long an effect lasts by selecting drug therapy, semi-permanent or permanent gene therapy. And with selective gene promoters, we can also program *when* and in *which situation* a gene will become active.

Designer Muscle

Every day, 14 people in the US are diagnosed with Amytrophic Lateral Sclerosis (ALS), better known as Lou Gehrig's disease. ALS is effectively a death sentence. On average, ALS patients live two to five years after diagnosis. Over those years, the nerve cells that control their muscles will die off. They'll progressively lose control of their body. The disease starts with twitching, cramping, and stiffness. Then the arms and legs become week. Speech becomes slurred. Chewing and swallowing become difficult. Eventually ALS sufferers will lose all voluntary muscle control. They'll become "locked in" - unable to move a muscle, unable to speak, but still perfectly aware of the world around them.

There's no know cure for ALS. The only approved treatment - a drug called riluzole - extends life by a few months at most. Recent animal studies, however, suggest that gene therapy could double the life expectancy of ALS patients and slow the progress of the disease.

In 2003, neuroscience professor Jeffrey Rothstein and his colleagues at Johns Hopkins gave mice suffering the equivalent of ALS extra copies of a gene called IGF-1. IGF-1 stands for insulin-like growth factor 1. It's a compound found in our bodies that, among other things, promotes muscle growth and repair. ALS normally progresses so quickly in mice that researchers can't test therapies *after* the mice show symptoms. Instead, they have to test techniques on mice before symptoms manifest, and then see if tested mice do any better than mice who don't receive the treatment.

The IGF-1 gene therapy worked differently. The researchers injected the mice after the symptoms first manifested. The mice responded amazingly well. They gained strength. Their nerve cells - the ones that normally die off in ALS - stayed healthy longer. And overall they lived about twice as long as the mice who didn't receive the gene therapy.

Encouraged by their results, Hopkins is now planning human trials of the therapy. If IGF-1 gene therapy protects humans against ALS, it will be the first truly effective weapon against the disease. Yet IGF-1 gene therapy may also find its way into other, non-medical uses.

In 1998, University of Pennsylvania biologist Lee Sweeney and his team delivered extra copies of the IGF-1 gene into the muscles of healthy mice. Their goal was to see if IGF-1 could slow the age-related loss of muscle that occurs in both mice and humans. In Sweeney's first experiment, he injected the gene into eighteen-month old mice, roughly the equivalent of fifty five year old humans. He then measured the mice's strength and muscle mass when the mice were twenty seven months old, the equivalent of eighty or older in humans.

To the researcher's surprise, the extra IGF-1 gene did more than slow down the age related decline in strength - it practically halted it. The twenty seven month old mice dosed with IGF-1 actually had a slight bit more muscle than they'd had at eighteen months. On some strength tests they showed only a tiny decline over the course of their lives, and on other strength tests they'd actually improved.

Intrigued by his results, Sweeney decided to study the effects in young adult rats, about the equivalent of twenty year old humans. His team compared three groups of rats: Genetically normal rats put on an intense strength training program for eight weeks, forced to climb up the sides of cages with weights tied to their bodies; Rats with an extra copy of IGF-1 who were completely sedentary; And rats who got both the extra IGF-1 gene *and* the strength training.

The first surprise was that sedentary rats with IGF-1 gained as much strength, about 16 percent, as the genetically normal rats on the rigorous exercise plan. What's more, the sedentary IGF-1 group actually gained *more* muscle mass. But the group that received both the IGF-1 gene therapy and the exercise blew the other two groups away, increasing their strength by 30 percent in the same time period. And two months after the study ended, the group that received only exercise had lost almost all of its strength gains, while the group that had received both IGF-1 and strength training had retained more than half of theirs.

Other groups have produced similar results with different genes. Another team led by Geoffrey Goldspink at the University of London gave mice extra copies of a variant of the IGF-1 gene called Mechano Growth Factor or MGF. The gene dosed mice showed a 20 percent increase in muscle mass and a 25 percent increase in strength, without any exercise at all.

Se-Jin Lee and colleagues at Johns Hopkins took a different approach. They bred a strain of mice that lacked a working copy of the gene for a hormone called myostatin. Myostatin normally suppresses the body's production of muscle. Lee's myostatin-lacking mice grew to have two to three times as much muscle as their genetically normal kin. And further study showed that they had 70 percent less body fat as well. Lee has gone on to find ways to reduce myostatin in adult mice. Knocking out myostatin in mice with the equivalent of muscular dystrophy helps them build muscle, so researchers are investigating the potential of the therapy in humans.

Gene therapy to boost muscle is likely to be even more popular than gene therapy to boost endurance. Unlike higher red blood cell counts, which can lead to heart attacks, there's no obvious health risk from large muscles.

Perhaps more importantly, the normal age-related decline in muscle has now been medicalized. In 1991 gerontologists William Evans and Irwin Rosenberg introduced the word *sarcopenia*, literally "vanishing flesh" to describe the phenomenon. Since then the word has appeared in dozens of journal articles, in publications of the Center for Disease Control, and so on. With its own fancy name, sarcopenia is on its way to becoming a first class disease. As a disease, it will be medically treated. Fifty-year-olds will have access to the treatment, and eventually it will spread to forty-, thirty-, and twenty-year-olds as well.

Cosmetic Gene Therapy

For most of us, though, the real appeal of muscle-boosting gene therapy may not be athletic, but cosmetic. Sweeney, himself a fit-looking thirty-something, sees this as a possible application:

> "Just say you'd like your pectoralis muscles to be a little larger because you want to look a little better at the beach. Just take a few injections of the virus, and a month later while you're watching television, your muscles have gotten bigger."

In fact, cosmetic uses may very well be more popular than sports enhancements. In the United States alone, more than 8 million people had some sort of cosmetic plastic surgery in 2001. That's one out of every thirty people or so. And in general, the easier a procedure is, the larger the population who'll try it out. Muscle boosting gene therapy may require just a single injection. The most comparable cosmetic procedure in the US today - the injection of Botox into facial to smooth out wrinkles – is also the most popular. In 2002, doctors in the US gave patients more than 1.6 million such injections.

The cosmetic applications of genetic technology go beyond boosting muscle. More than a dozen studies have shown that gene therapy can correct obesity in mice and rats. In 1996, researchers at the Howard Hughes Medical Institute in Chevy Chase Maryland gave mice extra copies of the gene for the hormone leptin. Leptin regulates bodyweight by controlling metabolism. Mice given a single gene therapy injection lost weight while eating just as much as undosed mice. Later studies produced similar results, and also demonstrated that the gene therapy protected the mice against diabetes. Diabetes and obesity are serious medical issues. Several research groups are working towards the first trials of leptin gene therapy in humans. Yet they also realize that anti-obesity gene therapy will have more than medical applications. As Sergei Zolotukhin, a professor at the University of Florida, noted, "This would be the couch potato's dream: You can eat what you want but stay lean." Given the billions of dollars consumers spend on diet products every year, it's almost a given that a one time injection that kept us thin while keeping us free to eat what we want would be hugely popular.

Another medical therapy with obvious cosmetic applications is the ability to change skin tone. Pale skin is a risk factor for skin cancer. An Australian company called EpiTan is conducting clinical trials of a drug that would give people a tan, and thus resistance to skin cancer, without the need for sunbathing. Gene therapy could darken skin permanently or semi-permanently.

In 2001, Heidi Scrabble at the University of Virginia created albino mice whose skin can change color. Scrabble and her colleagues gave the mice an extra copy the gene for the melanin. However, they used a promoter for the gene that would only activate in the presence of specially modified lactose. When the researchers fed the mice the modified lactose, their coats went from white to brown, and their eyes from pink to black. When they removed the dietary supplements, the mice went back to albino.

Not only did Scrabble and her team alter the color of the mice's skin genetically, they created a system for controlling those genes from the outside. In principle the same thing is possible in humans. You could receive an injection that gave you a tan. Or you could receive new genes that would allow you to change your skin tone - or other traits - via a pill. And as a long-lasting therapy against skin cancer it may just be medically viable.

Researchers are also hard at work on gene therapy as a technique to cure baldness. Robert Hoffman and colleagues at San Diego biotech company AntiCancer Inc. have shown that they can deliver new genes to the hair follicles of both mice and humans. Working with skin grafts rather than live animals, the researchers inserted genes for green fluorescent protein. The hair that grew out of the follicles glowed green under ultra-violet light. The study demonstrates that it's possible to inject new genes into hair follicles. Hoffman and his group are now searching for the genes responsible for baldness and grey hair. Once those genes are found, they plan to attempt to alter them in follicles, to see if they can preserve a full head of healthy looking hair.

Hoffman's research also points out the more eccentric cosmetic possibilities. There's no reason, for example, that gene therapy couldn't be used to deliver green fluorescence genes to human skin or hair. Such gene therapy would produce humans who glowed under blacklight. If done using a non-insertional vector, the effects would wear off in a few weeks or months, not much longer than a hair dye.

Indeed, other animals have a much wider variety of dyes in their bodies than we humans possess. Tropical birds and fishes are often adorned in bright reds, blues, and yellows. Those colors are the result of protein dyes created by the creatures' genes. In principle it would be possible to insert those genes into humans, producing brightly hued skin or hair.

Would society accept these sorts of uses of gene therapy? That's unlikely to happen anytime soon. In this case the technology is likely to outstrip public comfort. Yet the basic techniques of gene therapy are fairly simple. They require only a few thousand dollars of equipment and training available at any university. The starter materials - samples of viruses that can be used to ferry genes into the body - can be ordered from dozens of scientific supply companies. Thousands of labs in the US alone have the expertise and facilities to use them. Any number of graduate students or talented amateurs have the means to set up a basement lab in their home.

The ultimate limit on what people will do with gene therapy may not be science or technology, but simply human desire and imagination. If someone wants to make their skin glow for a few weeks, the technology will be there to allow it. In the meantime, it's likely that quite a few people will be interested in more mainstream cosmetic gene therapies, like permanent increases in muscle, or permanent changes to hair, eye, or skin color.

Risks and Benefits

While gene therapy has tremendous potential, it also has risks. Over the last few years the field has faced two serious setbacks. In September 1999, a young patient named Jesse Gelsinger died, apparently as a result of an experimental gene therapy he'd received for a rare disease called OTC. Then in 2002, a French gene therapy research team led by Alain Fischer and Marina Cavezzena-Calvo discovered that two of their patients, whom they'd just cured of another form of "bubble boy disease", had contracted leukemia, apparently as a result of the gene therapy.

The problem, in both cases, lies with the virus used to ferry the genes into the cells of the patients. Viral vectors have two major problems. The first is their effect on the immune system. Just as viruses have evolved to ferry genes into our bodies, our bodies have evolved immune systems to kill the viruses off. And when those immune systems are triggered, the side effects can be unpleasant, dangerous, or in rare cases deadly.

This appears to have been the cause of Jesse Gelsinger's death. Children with OTC lack an enzyme that helps break down the proteins in our diet. Without this enzyme, parts of the proteins combine with hydrogen and form ammonia. The ammonia spreads through the bloodstream and into the brain. In most children it's fatal - often within a month of birth, usually by age five.

At the age of eighteen, Jesse had a very mild case of OTC - some of his cells could still produce the enzyme. He probably didn't need the therapy himself, but volunteered for it in the hopes that the research would help save the lives of other children born with the disease.

Researchers inserted the functional OTC gene in an adenovirus - a virus that causes the common cold. The adenovirus had been stripped of the viral genes that cause the cell to build more virus copies. It had been used on thousands of patients before. It was considered safe, but something went wrong.

Perhaps because Jesse was already sick with a different virus, his immune system attacked the adenovirus vector. In the battle between his immune system and the several billion viruses injected into his body, red blood cells burst. The proteins in his blood cells were released into the rest of his body. Normally enzymes in his body would break down those proteins, but Jesse suffered from OTC - he lacked those enzymes. The gene therapy was supposed to fix that. Instead, the unprocessed proteins released by his burst red blood cells formed huge amounts of ammonia. His blood stopped clotting. His lungs failed. Ultimately his brain stopped working, and his family and doctors decided to turn off his life support.

The second problem, the one that struck the patient in the French study, is a lack of precision. Viruses insert the genes they carry into whatever spot in the host cell's genome they can reach. Usually this works out fine. The new gene is inserted in the middle of a fairly harmless or unimportant stretch of the genome. In some cases, however, the newly inserted gene can land in the middle of some important part of the genome - for example, it might appeal next to or on top of a gene responsible for fighting tumors.

As an analogy think of the genome like a book. The paragraphs in the book (the genes) are instructions for the cell, describing what proteins to make along with how and when to produce them. The gene added by gene therapy is a new paragraph, with its own instructions. Most of the pages of the book are so-called "junk DNA" which doesn't seem to serve any purpose. If you insert a new paragraph at some point in random, it will probably appear in one of these junk stretches. But occasionally it will appear in the middle of a page with important instructions, or even in the middle of another paragraph. That's when problems can arise.

This is what appears to have happened in the french study. The virus delivered it's genetic payload to a location on chromosome eleven known to be important in suppressing cancer. The new genetic material disrupted the function of the existing genes, allowing leukemia to develop unchecked.

Fortunately there appear to be ways around both the immune response and the imprecision of today's vectors. One of the most popular new techniques is to use adeno-associated virus, or AAV, as the vector. Around 80 percent of the world population is already infected with AAV, yet it appears to cause no disease. Perhaps because of this, the body's immune system hardly responds to AAV, making it a safer vector. It also tends to deposit it's genetic payload at a well known location, away from dangerous cancer causing genes, further increasing its safety.

In addition to safer viruses, researchers are working on a host of non-viral techniques to deliver genes to cells. The simplest is the 'naked DNA' technique – simply inject copies of the gene into the bloodstream, and some cells will take those genes up. Unfortunately this technique isn't very efficient. Only a tiny proportion of the inserted DNA makes it into cells. To improve on the method, researchers have packaged the DNA inside fatty molecules called lipids. They've learned to pack DNA into tiny 'nanoballs' small enough to fit through the pores on the surface of cells. They've used a process called electroporation to electrically widen pores in the surface of targeted cells. They've used tiny bubbles to deliver the DNA deep into the body and targeted ultrasound waves to burst the bubbles in precisely the right location to target specific organs. All told there are at least a dozen such techniques for delivering genes into the body without using traditional immune system-triggering viruses.

In addition, they're working on ways to target genetic changes to certain parts of the genome. Going back to the genome-as-book analogy, imagine being able to replace single words or letters at precise locations in the existing book, rather than adding new paragraphs on random pages. A number of researchers have made progress doing just that.

In 2002 a group at the University of Washington led by Roli Hirata and David Russell used an adeno-associated virus to ferry a piece of DNA into the cell. The inserted DNA was almost identical to an existing gene, and the cell's mechanisms detected this and replaced the current DNA sequence with the new, altered one the researchers had inserted. Instead of tacking on a new gene at a random location, Hirata and Russell had managed to replace a precisely targeted genetic sequence with a new one of their choosing. The team then went on to repeat the experiment with a second gene, suggesting that the technique could apply to a wide variety of DNA sequences in the human genome.

Geneticist Alan Lambowitz at the University of Texas and his team succeeded with another approach, based on the homing properties of bits of DNA called introns. Introns are just one of the many types of so called "junk DNA". Unlike other junk DNA, though, introns actually reside within our genes. Every gene is a long sequence of DNA. Within these DNA sequences are shorter stretches of exons and introns. When the cellular machinery that synthesizes proteins reads your DNA, it uses the exons to decide what kind of protein to build, and apparently throws away the introns. Introns almost certainly play some role in the cell, but the exact nature of that role isn't yet understood. What researchers discovered is that many introns can home in on specific locations in the genome and insert themselves. If the part of the intron's DNA sequence lines up with DNA already in the cell, it can trade places with the gene that's already there, and carry a bit of extra genetic information along with it.

Lambowitz and his team used this characteristic of introns to target them at specific genes. They demonstrated that in human cells they could direct the introns to a chosen location. In bacteria they've used this ability to disrupt dangerous genes. Lambowitz believes they'll also be able to use introns to ferry new genes to a location of their choice. "People who do gene therapy use retroviruses as vectors, which are inserted randomly so there is no control over where they are going. It's a shotgun approach. The intron is more like a rifle shot. You can make it go to the site of your choosing. You can knock out a gene that causes cancer or introduce a tumor-suppressor gene or target a safe site where you will not impair the function of anything else," he explains.

All of these techniques are experimental. There's no guarantee that any one of them will work. As a science, gene therapy itself is young – less than twenty years old. Its promise lies in what it's already shown us – that it's possible to alter human genes, and that by doing so we can cure disease.

FOR ALL OF THE PROMISE of gene therapy to cure disease and sculpt our bodies, there are also several frequently voiced concerns. The most frequent are that altering human genes is unnatural or that gene therapy is unethical because it's not yet safe.

The first arguments – that gene therapy is unnatural – rings hollow. It may not be natural to insert new genes into the human body, but the same argument could be leveled - and has been - at almost any medical technique. In 1798, Edward Jenner's vaccine against smallpox was denounced as unnatural and immoral in newspaper editorials and from church pulpits. According to historian Andrew Dixon White, anti-vaccination societies condemned vaccination as "bidding defiance to Heaven itself, even to the will of God," and declared that "the law of God prohibits the practice."

The use of anesthetics during childbirth was considered "unnatural" and "meddling in a natural process" until Queen Victoria championed it in the mid 1800s. It's not natural to take blood out of one person and give it to another, yet transfusions are a well accepted part of medicine. It's not natural to transplant organs, but we've done that with hearts, livers, kidneys, and more. The natural thing to do would have been to let Ashanti DeSilva die. Thirty years ago, she would have died within the first few years of life, simply because we lacked the knowledge to keep her alive. Even peg-ADA, the drug that kept her alive up until her gene therapy is an unnatural product – a synthetic version of an enzyme normally created in the body.

New and unfamiliar medical techniques are often seen as unnatural. Yet with time and usage come familiarity and thus acceptance. By the early 1800's the smallpox vaccine was in widespread use around the globe. Within years of Queen Victoria using chloroform at childbirth, women throughout England were doing the same. Gene therapy may seem threatening today, but as it becomes more effective, as it's used to treat more and more diseases, it will seem less alien and more like any other medical technique we're accustomed to.

The safety argument deserves more consideration. As the cases we discussed a few pages ago make clear, gene therapy still comes with substantial risks. The same is true of every technology we'll discuss in this book. Any attempt to tinker with the human mind or body in a new way is going to have possible complications. Sometimes, as in the case of Jesse Gelsinger, those complications can kill.

Society could eliminate this risk by simply banning experimental medical techniques. Yet that would end medical progress. Every medical technique has a first human subject. Scientists and doctors run thousands of trials with animals or cells in a dish to try to ensure safety. Many techniques wash out at this stage. Only one tenth of one percent of candidate drugs will perform sufficiently well in lab and animal experiments to be allowed to reach human trials. Eventually, though, if a medical technique that looks safe and effective in the lab is going to benefit real people, it needs to be tried on a first set of patients.

These first patients are taking risks. Yet that risk is relative to their current circumstances. The first people to go through gene therapy - or any possible enhancement technology - will be those who are extremely ill. They'll be using the technology not to increase their abilities to superhuman norms, but simply to live, or to regain abilities that disease or injury have taken away from them. Given the choice between a disease one knows or a possible cure with unknown side effects, many sick and injured will chose the latter. That's the position Ashanti DeSilva's parents were put in. Watch their daughter's almost certain decline and death from ADA-deficiency? Or put their hope in an untested therapy? In the DeSilva's case, the gamble succeeded. Ashanti is alive today because her parents were willing to take the sliver of hope offered by a new technique.

Yet there are also times when gambles fail. The first heart transplant patient lived just eighteen days after the operation. Barney Clark, the first man to receive an artificial heart, died one hundred and twelve days after the implant of the device. Many of the first patients for other groundbreaking medical techniques died either as a result of complications, or simply because the therapy failed. Every new medical technique has had its share of problems. Yet these problems aren't in vain.

The experimental gene therapy that killed Jesse Gelsinger taught researchers about the risks of immune response to their viral gene vectors, and spurred research into other, safer techniques. Gelsinger won't benefit from those lessens, but the next generation of children born with OTC might. Jesse seems to have realized as much. Before his therapy he told a friend, "What's the worst that can happen to me? I die, and it's for the babies."

We'll see this same pattern again and again throughout this book. Every technology we'll talk about can be used to heal or to enhance. And in every case the early adopters are those who have the most to gain and the least to lose. In some cases those first transformed men and women have benefited tremendously. In others they've suffered side effects. Yet in every case they've helped teach us more about the human mind and body - leading to better, safer techniques that can heal millions of other sick and injured.

Ironically, one of the most obvious responses to safety concerns about enhancements - to ban the enhancement techniques - is counterproductive. Enhancement techniques are likely to be quite popular. Consider some of the precedents: In addition to the more than 8 million cosmetic plastic surgery cases they underwent, U.S. consumers in 2002 spent $17 billion on sports supplements and herbal products intended to improve health or boost mental or physical function. Most of these supplements have little or no impact, yet they're incredibly popular. When truly effective mental and physical enhancements are available, they'll tap into a large existing demand.

Bans on highly in-demand goods and services don't seem to eliminate the market for such things. They simply move it underground. That's what happened in the 1920s when Prohibition in the US made alcohol illegal. It's also the case today in the US with the War on Drugs. Neither effort stopped the flow of the banned good. They simply created black markets for the products people sought.

In a black market, safety suffers. There are no regulators to enforce safety standards. There's no threat of liability for botched services or procedures. It's difficult to perform long-term safety studies to spot emerging problems.

For example, illegal drugs sold on the street are often not what they're claimed to be. In 1999, drug safety researchers determined that the majority of emergency room visits in Oakland California that were attributed to use of the drug MDMA ("ecstasy") were actually the result of dangerously high doses of DXM, an ingredient found in cough syrup, pressed into tablets and falsely sold as ecstasy. In a regulated market, this seldom happens. Inspectors and safety checks largely guarantee that the consumer gets what he or she is paying for.

Another example is the effect that bans on abortion have on safety. In countries where abortion is legal, only one out of every hundred thousand abortions leads to death. In countries where abortion is illegal, the death rate can be seven hundred times higher. The World Health Organization estimates that worldwide, one out of every eight maternal deaths is due to an unsafe abortion - most of those in areas where abortions are illegal. When a service is made illegal, many of the safety standards regulators insist on go by the wayside. If safety of enhancement techniques is a concern, it makes more sense to regulate them than ban them.

That regulation must also take into account the use of these techniques as enhancements, not only as medical procedures. The cyclists who died from overdoses of EPO died because they lacked good information on how much was safe. The literature around EPO and the safety tests that EPO products had to go through prior to regulatory approval were focused on the medical uses of the drug. The studies involved animals and patients with anemia, not healthy men and women, let alone athletes at the top of their game. Banning the enhancement use of EPO didn't stop cyclists from using it – it just robbed them of data that might have saved their lives. A truly effective safety system would acknowledge that people will use medical products as enhancement, and establish safety for exactly those uses.

In just a few decades we've gone from the first tinkering with human genes to the discovery of dozens of techniques that could precisely alter the human genome. Those techniques give us the power to cure diseases or to enhance and sculpt our bodies. This new control over our genes promises to enhance our quality of life as dramatically as the medical discoveries of the past century.

Chapter 2 – Designer Minds

One out of every ten people aged sixty-five or older has Alzheimer's disease. Nearly half of those over age eighty-five have it. Worldwide there are more than 12 million sufferers of the syndrome. By 2025 that number is expected to almost double, to 22 million. In the US alone consumers and insurance companies spend over $100 billion on the disease each year, most of it on medical and nursing care.

Alzheimer's devastates individuals and families. As memory and competence fade, the Alzheimer's patients become dependent on their spouses, their children, or whomever can be found to care for them. For the caregivers, it's a long slow process of watching a loved one fade away, compounded by the often crushing financial burden of caring for their relative.

Small wonder that so much effort is being put into finding a cure. The National Institute of Aging spends half of its billion dollar annual budget on research into Alzheimer's. Dozens of private companies and university labs funded by other sources are looking at the disease as well.

One of the more promising approaches is gene therapy. On April 5th of 2001, Mark Tuszynski and other researchers at the University of California in San Diego (UCSD) implanted genetically modified neurons into the brain of a 60 year old woman suffering from Alzheimer's. The nerve cells were modified to have extra copies of the gene for Nerve Growth Factor, or NGF.

NGF is a chemical that triggers the growth of neurons. Levels of NGF slowly drop in the brain as we age. Perhaps as a result, neurons in aging brains tend to shrivel up and die. In particular, the axons of neurons, the long fibers that extend out of them and are used to signal other neurons, tend to shrink.

In a series of earlier studies, Tuszynski had found that adding extra copies of the NGF gene could protect the brains of mice and monkeys against this age-related shrinkage. What's more, it could actually *restore* the neurons in old brains to their youthful size, shape, and activity. Persuaded by Tuszynski's findings in animals, the Food and Drug Administration approved his proposal for human trials - one of the few human gene therapy trials to target the brain.

After following his patients for years, Tuszynski and his team found that the NGF gene therapy, while it didn't cure Alzheimer's, slowed the rate of onset by a factor of three - making it one of the most effective tools against Alzheimer's yet developed. If larger trials find the same thing, the therapy could be used on a large scale within the next several years.

Yet NGF may do more than merely restore lost function. In 2000, Howard Federoff and colleagues at the University of Rochester discovered that NGF gene therapy could *improve* the learning and memory of normal mice. Federoff compared normal mice to mice genetically engineered to produce more NGF in one part of their brain. By the third trial of a maze-navigation test, mice with extra levels of NGF could learn the maze about 60 percent faster than normal mice.

In another test, NGF made an even bigger difference among mice who were raised in a more intellectually challenging environment - one with more toys, obstacles, puzzles, and tests for them to complete. NGF-enhanced mice raised in the rich environment learned the maze *four times as fast* as normal mice in the rich environment, and more than *five times* as fast as normal mice raised in ordinary cages.

Learning to navigate a maze is one thing. Human learning - with its focus on language, mathematics, abstract ideas, and interpersonal relations - is a more complex thing. Yet the NGF-enhanced mice illustrate the power of genes to enhance learning and memory, and a number of other studies with different genes have produced animals with just as big a mental edge.

Princeton biologist Joe Tsien's *Doogie* mice are probably the most famous of these "smart rodents". Named after TV's child prodigy Doogie Houser M.D., the mice were featured in *Time* magazine and the David Letterman show. On a variety of learning tests, they perform better than normal mice - in some tests as much as five times better.

To understand what Tsien did we have to understand a bit about how the brain works. Your brain has around a hundred *billion* neurons. That's fifteen times as many neurons as there are people living on this planet, and on average each of those neurons each connect to about one thousand other neurons. So you have around one hundred *trillion* of those connections between neurons - what scientists call *synapses*.

Synapses are where messages are conveyed from one neuron to another. Messages generally come in the form of individual molecules called *neurotransmitters*. One neuron, the pre-synaptic one, fires. It releases a flood of these molecules. The message is conveyed when these molecules bump into the matching *receptors* on the *post*-synaptic neuron, the one on the listening end. Receptors are like 3-dimensional puzzle pieces. They're shaped so that only certain neurotransmitter molecules will fit well inside them. So for the signal to be sent, the neurotransmitters released from the pre-synaptic neuron have to be the right kind to dock snuggly with the receptors on the post-synaptic neuron.

The most common kind of neurotransmitter molecule is glutamate. Neurons that send signals with glutamate are found in just about every nook and cranny of your brain. Tsien and others have focused mostly on glutamate-using neurons in the *hippocampus*, the part of the brain involved in long term memory.

Neurons in the hippocampus, as in most of the rest of your brain, experience something called long-term potentiation or LTP. When two neurons fire in synch, the connection between them tends to get stronger. It's *potentiated*. And this potentiation, this strengthening, can last for weeks or months or years. Hence the phrase '*long term potentiation*'.

Tsien knew that as humans age, LTP becomes less and less frequent. It takes more signals between neurons to convince them to strengthen their connections. He also knew that as we age our brains produce less of a protein called NR2B, which is a crucial part of the NMDA receptor - one of receptors found in the hippocampus and which responds to signals from glutamate. Tsien reasoned that these two phenomena could be related, and could explain why young people have better memories than older people.

That led him to genetically engineer a strain of mice with extra copies of the NR2B gene - the *Doogie* mice.

The results surprised even Tsien. The Doogie mice didn't just hang onto their memories in old age. They actually formed memories more quickly - *learned things* more quickly - at any age. After being shown an object for five minutes, Doogie mice remember it five times longer than normal mice. When placed in a water filled maze, Doogie mice learn how to find the hidden platform they can remain dry on twice as quickly as normal mice. When given an electrical shock in a room, the Doogie mice are about 30 percent more likely to remember it ten days later. If researchers turn off the electrical current in the room, Doogie mice figure that out twice as quickly as normal mice. This last point is important. Not only do the Doogie mice *learn* more quickly than normal mice, they also *unlearn* more quickly.

Faster learning and better memories aren't necessarily free blessings, though. The NMDA receptor is involved in the response to addictive drugs like cocaine and heroin. NMDA is also involved in stroke. That raises the possibility that people taking NMDA-affecting drugs or gene therapy could be at greater risk of drug addiction or stroke. Later studies also suggested that Tsien's mice feel major pain longer than normal mice. After being injected with a noxious substance in a paw, the *Doogie* mice lick at it longer than other mice.

Despite these possible problems, interest in applying Tsien's research to humans has been intense. Indeed, in 2003 researchers discovered that Alzheimer's patients had lower than normal activity of the NR2B gene, suggesting that Tsien's research could be directly applicable to treating the disease. In addition, a number of other companies have been formed with the intent of creating memory-improving drugs based on related parts of the memory system.

Other groups are working on genes that affect what goes on inside the cell *after* the NMDA receptor is activated. In the 1980s Nobel-prize winner Eric Kandel discovered that a chemical called CREB is involved in the changes inside a neuron that are involved in building and strengthening a synapse.

In 1994 a competing researcher named Tim Tully built on Kandel's work. Tully and a colleague at Cold Spring Harbor Laboratory in New York added a gene to fruit flies to increase the amount of CREB in their brains. The resulting fruit flies had astoundingly good memories. For example, they could lean to associate a smell with an electrical shock in one repetition, while normal flies took 10 tries to learn the same thing. Interestingly enough, the genetically engineered flies didn't do any *better* after more repetitions. After 10 repetitions the normal and engineered flies performed the same. But the flies with more CREB got to the same level of performance as the normal flies in one tenth the time.

Experiments in mice and sea slugs have shown the same effect as in fruit flies. Animals genetically engineered to have higher levels of CREB learn things faster. The genetically engineered animals also seem perfectly healthy. Other than their boosted learning, they behave just like other members of their species.

In 1998, Kandel leapt ahead when he showed that a drug that boosted the amount of CREB in the brain could improve long term memory in mice. Kandel found that injecting mice with Rolipram, an anti-depressant that indirectly raises levels of CREB, could boost mouse memories. In tests of their memory 24 hours after training, Rolipram-injected mice did about 50 percent better than mice injected with saline solution. Tully did his own experiments and found that mice injected with Rolipram could learn things in 2 trials that other mice required 5 trials to get right.

Emboldened by success, Tully and Kandel went on to found competing companies. Rolipram, as it turns out, is not a good drug option for most people. Its side effects include nausea and vomiting. Kandel's company, Memory Pharmaceuticals, and Tully's company, Helicon Therapeutics, are both working on new drugs that work similarly to Rolipram but which can be taken as pills instead of injections, and which don't have the nasty side effects. Both drugs - MEM 1414 and HT-0712 - are just entering human trials.

In total, at least a dozen companies in the U.S. alone are pursuing some sort of memory enhancer drug today. The motivation is a large and glittering market. 4.5 million of the world's Alzheimer's patients live in the U.S. In addition, there are another 4 million or so Americans with mild cognitive impairment or MCI, a syndrome the FDA now recognizes as a treatable disease. Every year around 10 to 15 percent of MCI patients degenerate to the point of full blown Alzheimer's disease. Most recently, the FDA has recognized "age associated memory impairment" or AAMI as a treatable disease. Anyone over age 50 who is in the bottom 16 percent of his or her age group in memory tests is defined as having AAMI. That's 12 million people in the US.

The market may be larger yet. Tully notes that there are now almost 80 million people aged 50 or older in the US, and that this age group will double by 2030. And UC Irvine's James McGaugh, a critic of memory-enhancing drugs, told *Forbes* magazine that "Drug companies won't tell you this, but they are really running for the market of nonimpaired people—the 44-year-old salesman trying to remember the names of his customers". Axel Unterbeck, President and Chief Science Officer of Kandel's Memory Pharmaceuticals concurs, saying that, "If it is safe, the market is incalculable."

For his part, Tim Tully notes that eventually "memory enhancers could become 'lifestyle' drugs to be used by anyone interested in learning a second language, in playing a musical instrument or in studying for an exam."

Ultimately, individual people will judge the pros and cons of memory enhancers. That process is starting already, with the first few human trials of the current round of drugs. If those trials are successful, more studies will follow. Eventually, elderly patients may start using the drugs. If they report satisfaction, usage will spread to a wider population. If the benefits of the drugs outweigh their cost and side-effects, they may be used in just the way that Tully suggests – by adults who want to learn to play the piano, or pick up a new language, or by students trying to improve their grades.

On the other hand, the current drugs could fail in human trials. The most likely way for them to fail would be unpleasant side-effects. If, like Rolipram, they make their users physically ill, they're unlikely to become popular. But whether or not the current batch of drugs succeeds in the market, the last decade has taught us a tremendous amount about the molecular basis of learning and memory. Sooner or later, someone will figure out how to make a drug that improves human learning without serious side effects. As Tully told *Forbes*, "It's not an 'if' – It's a 'when'".

Smart Drugs Are Here Already

Both smart drugs and smart genes may seem farfetched, but in fact they're already so entrenched in our culture that we hardly notice them. Hundreds of millions of people already use some sort of drug to improve their memory or attention. Caffeine and nicotine are so familiar to us that we don't think of them as performance enhancers, but that's exactly what they are. Every time you have a coffee or caffeinated soft drink to stay awake while driving, you're engaging in a mental enhancement – pushing your alertness beyond its normal human endurance.

Caffeine and nicotine slightly improve reaction time and attention. Nicotine also improves long term memory and performance on problem solving tests. As any smoker will tell you, nicotine helps people concentrate. Stanford researchers have even found that well-rested pilots given nicotine before a flight simulator test perform better than control subjects. The biggest improvement is in complex and attention-demanding tasks like landing the plane and responding to simulated emergencies.

While nicotine usage is dropping, the socially-sanctioned use of other, more powerful cognition enhancers is on the rise. In 1999, physicians in the US wrote more than 15 million prescriptions for stimulant drugs Adderall (amphetamine), Ritalin (methylphenidate, a close cousin of amphetamine), and their clones. Those prescriptions were almost all for Attention Deficit Hyperactivity Disorder, or ADHD.

Studies suggest that ADHD is diagnosed too often. At the same time, it's clear that most children and adults who are diagnosed with the disease respond well to the stimulant drugs that are prescribed for it. On average, Ritalin and other drugs improve attention, memory, test scores, and grades. Among the most severe ADHD children, they reduce the risk of legal trouble, of dropping out of school, and even of later substance abuse.

What's less appreciated is that drugs like Adderall and Ritalin improve attention and memory in *almost everyone who takes them*, whether they are diagnosed with ADHD or not. NIH researcher Judith Rapoport has found that ADHD drugs give the same mental boosts to normal children and adults as they do to hyperactive ones. As psychologist and professor Ken Livingston wrote in *The Public Interest*, "even if you have never been diagnosed as having a problem paying attention, many of these drugs will improve your focus and performance." Lawrence Diller, author of *Running on Ritalin*, agrees. He calls Ritalin a "universal performance enhancer" and notes that when placed on the drug normal children develop better-than-normal mental memory and attention.

Another drug that's recently caught the public eye is Modafinil. Better known by its trade name Provigil, Modafinil is a French drug approved for use in treating narcolepsy. Narcoleptics are prone to sudden and intense drowsiness and even uncontrollable bouts of sleep. Modafinil gives narcoleptics the ability to remain alert and awake throughout the day. It can also be used to stay awake for extraordinary lengths of time – up to several days in a row. While awake, modafinil users suffer less jitteriness and fewer other side effects than people given stimulants like caffeine or amphetamine.

Unlike stimulants, people under the influence of modafinil can sleep when they choose to. When they do eventually sleep, they suffer less of a "rebound" than people kept awake for the same amount of time on amphetamine or nothing at all. For these reasons, the US military is testing its use for pilots and other soldiers as part of its as part of the Continuous Assisted Performance program.

The other ubiquitous mental enhancer of our time is the anti-depressant. Almost 38 million people in the world today are prescribed Prozac, and at least as many are taking some closely related drug. We don't usually think of anti-depressants as enhancers. After all, their real goal is to restore someone to a more "normal" state of mind. The reality is that Prozac and its clones affect both the depressed and the non depressed.

For example, a 1998 study in by Victor Reus and Brian Knutson at the University of California, San Francisco found that that the anti-depressant Paxil made already perfectly happy people more socially adept in challenging group situations. In group problem solving studies, people given Paxil were rated as more effective partners by judges who didn't know who'd been given the drug and who hand't. Another study that year showed that Paxil reduces anger in non-depressed individuals. Animal studies have shown that monkeys given drugs like Prozac are more likely to rise to the top of their social hierarchies in times of turmoil.

Antidepressants and ADHD drugs have their downsides. Prozac interferes with orgasms. Ritalin and Adderall can produce jitteriness, headaches, irritability, or increased blood pressure. In extreme cases they can turn a formerly creative, adventurous child into a listless or depressed one.

Most seriously, Ritalin, Adderall, and their ilk can easily be abused, even to the point of addiction. Adderall, the second most prescribed drug for ADHD, is just a trade name for pure amphetamine. Ritalin is closely related. Both have the same general effect on the brain as methamphetamine ('crystal meth') or cocaine. All four drugs mimic the effects of the neurotransmitter dopamine, which your brain normally releases in response to extreme pleasure or to reinforce positive behaviors. When you take a dopamine-mimicking drug, the brain reinforces the action of taking the drug. Frequent use at high doses increases the reinforcement.

In the long term, gene therapy may provide a long lasting or permanent alternative. Researchers have already linked a mutation of the gene responsible for one kind of dopamine receptor to a heightened risk of ADHD. They've identified two genes on chromosome 13, called G30 and G72, that increase risk of bipolar disorder. And they're known for years that mutations in a gene for the serotonin transporter, the part of the neuron that sucks serotonin back in and recycles it, are associated with depression, anxiety, and binge drinking.

At some point, doctors may use gene therapy to treat depression, ADHD, or bipoloar disorder. So far that hasn't been tried, but gene therapy experiments in both animals and humans have shown some interesting results in other cognitive areas.

For cancer sufferers, amputees, and burn victims, chronic pain is an immense problem. The drugs that are most effective in controlling that pain - opiates like morphine - have serious side effects. They dehydrate patients, make them drowsy and constipated, and can lead to physical addiction. A team at the University of Pittsburgh led by David Fink is trying to find ways around that. In 2005 they're set to launch the first human trials of gene therapy to control pain. In previous studies, Fink and his colleagues gave mice extra copies of the gene for proenkaphalin, a pain-killing molecule naturally found in the body. The mice whose bodies produced more proenkaphalin showed a higher pain threshold than genetically normal mice. What's more, they didn't show any sign of the side effects that opiate drugs cause. Numerous other groups have seen similar results. Fink hopes that the research will lead to techniques to control chronic pain in patients.

Another intriguing area is gene therapy for alcoholism. Researchers have known for years that low levels of a certain type of receptor for the neurotransmitter dopamine are associated with increased risk of alcohol addiction. In 2001, Panayotis Thanos and Nora Volkow at Brookhaven National Labs tried to see if they could correct that through gene therapy. The researchers delivered extra copies of the gene for this receptor type (the D2 receptor) into the brains of rats who had normal levels of D2, but who'd become addicted to alcohol anyway. The neurons in the rat brains grew more of the dopamine receptor. In the days that followed, the most profoundly addicted rats cut their drinking habits by two thirds.

Gene therapy might even affect our more personal relationships. In 2001, Larry Young and colleagues at Emory University in Atlanta showed that new genes could change mating behavior. The researchers gave male prairie voles gene therapy to increase the number of receptors in their brain for the hormone vasopressin. The voles that received gene therapy bonded to female voles more quickly than normal, even without the usual prerequisite of sex. They're also more friendly with other males and less likely to take risks.

Researchers already know that vasopressin is involved in human romantic bonding. Along with oxytocin, it's one of the brain chemicals associated whose levels increase over the course of a romance. Something as simple as a kiss can release more vasopressin into your blood stream. The finding that changing the genes for vasopressin can change the mating behavior of other mammals dovetails nicely with this. And it suggests that men and women could chemically alter their relationships if they so chose - either temporarily with drugs or permanently with gene therapy. Someday, for instance, clinics might offer couples gene therapy to strengthen and extend the bond of their relationship.

There are a whole range of other genes that have been associated with behavior. - genes that we might be able to use to sculpt our own personalities For example, variations in the gene for a brain chemical called monoamine oxidase A (MAOA) are associated with thrill-seeking and novelty-seeking behavior. We know that a particular kind of serotonin receptor, the 5HT2-C receptor, is involved in a variety of spiritual experiences. There's the still-controversial claim that a gene somewhere in the Xq28 region of the X chromosome increases the likelihood of homosexuality. At the other end of the spectrum, researchers Karen Carver and Richard Udry at the University of North Carolina have calculated that genes account for 26 percent of the variation in strength of religious belief.

In the next few decades this accumulated knowledge base could be used to create new drugs that sculpt or alter any aspect of human behavior: infatuation, pair bonding, empathy, appetite, spirituality, thrill seeking, arousal, even sexual orientation. If and when gene therapy in the brain is ever feasible, it will be possible to opt for permanent or semi-permanent alterations of personality. As a species we're in the process of figuring out what makes us tick. The next step - a step tens of millions of consumers will pay for - is to turn that knowledge into products that allow us to sculpt our own minds.

Neurotechnology and Society

Greater power over the human brain and mind will have profound effects on society. New mind-altering drugs and gene therapies hold promise in treating Alzheimer's disease, age-related decline, chronic pain, and more. Smart drugs and smart genes alone may improve the quality of life of tens of millions of people who currently struggle with mental illness.

The benefit in social and economic terms may be just as large. Americans alone spend almost $100 billion per year on long-term care for Alzheimer's disease patients. According to the National Institutes of Mental Health, other mental illnesses including depression cost the U.S. almost $300 billion per year in lost worker productivity, medical care, secondary illnesses, and so. Reducing the incidence of Alzheimer's, depression, or other major mental diseases would save tens of billions of dollars in the U.S. alone. *Improving* the learning speed and attention of mentally healthy people may reap productivity gains of hundreds of billions of dollars more.

People with better memories and quicker minds will earn more money and produce more for others. Any technique that increases the human ability to learn, to think, or to communicate is going to produce economic returns. It will increase our ability to solve problems, to make scientific breakthroughs, to build better products, and so on. Scientists that can learn more quickly will be better able to stay abreast of their fields. Doctors and nurses that can stay alert longer will make fewer errors in treating patients. Smarter engineers will produce better products to improve our lives. Smarter programmers will make better software. Smarter architects will design better buildings. Smarter biologists will come up with new medicines. Overall, society will become richer.

This isn't just speculation - it's supported by years of study. Around the world, the correlation between the average IQ of a country and its GDP is an incredibly high 0.76. The correlation between average IQ and rate of economic growth over the last quarter of the 20th century was 0.64. According to Harvard professor Robert Barro, a single standard deviation gain in science and math test scores raises economic growth rates by around 1 percent. In the US that's a gain of $1 billion per year to the economy for every point gained on the SAT.

Another study by the British Association of University Teachers found that every additional year of education across the population adds around half a percentage point to annual economic growth rates, and that the percentage of a country's GDP that it spends on education has around a 0.5 correlation with its wealth. Stronger memories that reduce the amount of time needed to learn a subject will make that educational investment more effective. Smart-drug or smart-gene enhanced students will be able to learn more with less time in school.

That money isn't an abstract - it's real improvements to the quality of our lives. Better homes, safer cars, more effective medicines. Improving the human mind improves all the products of the human mind. Any technique that improves on the human intellect will pay dividends throughout society.

This has competitive consequences as well. It's possible that in the near future some countries will have banned mind-enhancing drugs and other enhancements while other countries have embraced them. For example, the United States and Western Europe might prohibit any kind of non-therapeutic mental enhancement while in China, India, and much of Asia they become commonplace.

This is not at all far-fetched. Asian countries are in general more friendly to the idea of human enhancement than western countries. Polls on attitudes towards genetic engineering, for instance, find that while only about 20 percent of Americans approve of the use of genetic techniques to create desired traits in children, 63 percent of the population of India approve, as do a whopping 83 percent of those polled in Thailand. While those numbers are specific to genetic engineering, they suggest something of the fundamental attitude towards enhancement. If drugs or gene therapies that boost mental performance are outlawed in the United States, they may still move forward in other countries, and the nations that embrace them will gain a competitive advantage over those that do not.

With the benefits of enhancement come some potential drawbacks. Deeper knowledge of the brain will make it easier to develop new recreational drugs. Many of those will be quite harmless, even enriching. Others may be addictive and dangerous if used without precaution.

The power to control our emotional states also raises questions for our sense of identity. Once the technology is mature, some people will choose to alter their personalities in ways that their friends and loved ones don't understand. Rare individuals may seek out ways to increase their levels of aggression, or violence, or hatred, or to suppress their capacity for guilt, remorse, and empathy.

Society isn't likely to sanction the more radical of these changes, yet as the technology matures it will be easier and easier to put to use on the black market. There's no reason why, in the long run, splicing a gene into a new vector will be any harder than cooking up a batch of illegal drugs. And as we can see, there are plenty of chemists with the skill and motivation to do the latter. Where there is demand for an illicit mental alteration, there will be supply.

Even with less drastic changes, the ability to alter our own minds will bring us face to face with some deep and perplexing questions about who we are. Am I still the same person if an injection changes parts of my personality? If I go from shy and reserved to being a risk-taker and thrill-seeker? What if I always wanted those traits for myself? What if my friends no longer like me, or I no longer like them? If my wife and I take injections to increase the intensity of our neurochemical bond, what does that say about our relationship? Has our love been deepened or demeaned?

These questions could as easily be asked without any new technology. Am I the same person under the influence of alcohol or other drugs? Am I the same person I was 10 years ago? If I learn a new language, move overseas, make new friends, get a new job, or change my religion, am I still the same person I was before?

Neurotechnology doesn't radically alter the nature of identity – it just brings some of the limitations of the idea into starker relief. The reality is that we're constantly changing. Every experience we have alters us – intellectually, emotionally, neurobiologically. As you read this book electrochemical impulses are racing between the neurons in your brain. In those neurons, genes are being turned on and off. Dendrites are stretching out to close the gap between neurons firing in sync. Other connections are weakening and separating. Your brain will never be the same as it was before you read this page.

At the psychological level as well, permanent changes are being made. Memories are being formed. Ideas are being triggered. Older thoughts are resurfacing. Your future behavior is being altered slightly. What you've read and your reactions to it may affect your choices in other books or movies or conversations tomorrow or next week or next year.

These changes are all happening because of a choice you made. You chose to pick up this book. The other choices you make – what to read, what to watch, where to go, who to spend time with – affect your mind and brain as well. Some choices – what to study in school, where to live, what job to take – can have a profound impact for years to come. To a large extent, the choices you make today affect who you'll be in the future.

Neurotechnology differs only by degree. It will make this power to choose who we are more obvious and immediate. That, in turn, will make individual responsibility for our behaviors and identities more clear. In a world where we can sculpt our own emotions and personalities, people will no longer be able to say "I can't help it, that's just the way I am." Undoubtedly that transition will cause problems. Yet in the end it seems a fundamentally positive development – an increase in our power to determine who we are, and a greater responsibility for who we become.

Chapter 3 – Created Equal

In a 1997 commencement address at Morgan State University, Bill Clinton said "science and its benefits must be directed toward making life better for all Americans--never just a privileged few. Its opportunities and benefits should be available to all."

Clinton's statement could be stronger. For science to be just, one could argue, its benefits must be available to everyone around the world, not just those in rich western countries.

The power to sculpt our minds and bodies is an outgrowth of our efforts to heal. The same is true of technologies that can slow human aging, alter the genes of our children, and link the best abilities of computers with our brains. All of these enhancements emerge from medical research, so we can turn to medicine for an assessment of the likely economics.

The history of medical technology and its prices suggest two reasons to believe that enhancement technologies could become cheap enough for most of the world to benefit from. First, while most enhancement techniques we'll discuss are expensive to develop and test, they're cheap to manufacture. That means that after the initial research and development is done, most of the money has been spent. From there on out, the cost of the product tends to drop. Like almost any new technology, human enhancement techniques will at first be available only to the rich, but over time they'll become more affordable, allowing more and more people to reap their benefits.

Second, spending on biotech enhancements is likely to obey a law of diminishing returns. That is to say that someone who can spend ten times as much on an enhancement isn't going to get ten times the result. They're likely to get much less than that. The biggest gains are going to come from relatively small amounts of spending. Everything beyond that will produce only small refinements.

Let's focus first on price. Consider the cost of a prescription drug. When you walk into a pharmacy and fill a prescription, what are you really paying for? To you it may seem that you're paying for the substance contained in those pills or tablets. After all, what you care about is putting that chemical in your body and enjoying the beneficial effects – a reduction in pain, the elimination of some disease, or an alleviation of depression - but in many ways, what you're really paying for is information.

Drugs cost almost nothing to manufacture. Usually once researchers figure out that a certain molecule causes a certain effect in the body, almost any laboratory can synthesize the substance at a reasonable cost. And the more of it they make at once, the lower that cost goes.

The real cost is in the research and development that leads to a new drug - thousands of experiments, years of work, dozens of other candidates thrown away, and long and costly safety trials to get approval. According to a 2000 study from Tufts University, the average cost of developing and approving a new pharmaceutical is now close to a billion dollars.

Pharmaceutical companies don't make this investment out of altruism. They're motivated by the potential profits, and those profits are protected by drug patents. While the patents on a new drug are in effect, only the company that holds those patents can determine who is allowed to manufacture the drug. That company can charge as much as it thinks customers are willing to pay.

Drug patents last for 20 years after the invention of a drug. Often it takes a decade or more for a drug to reach the market after its invention, so for practical purposes a patent preserves a monopoly for the manufacturer for a decade or less. Once the patents expire other companies can manufacture and sell generic versions of the drug, without having to spend the huge amounts to develop it again from scratch. When multiple companies sell the same drug, they're forced to compete on the basis of price, driving the cost down for consumers. The greater the demand, the lower the price goes. In addition, higher demand leads to greater economies of scale in production and increased price competition as more companies manufacture the drug. According to a 1998 Congressional Budget Office study, drugs whose patents have expired and which are manufactured by more than 20 companies cost on average one fifth of what still patent-protected drugs do. We can see this in the graph below.

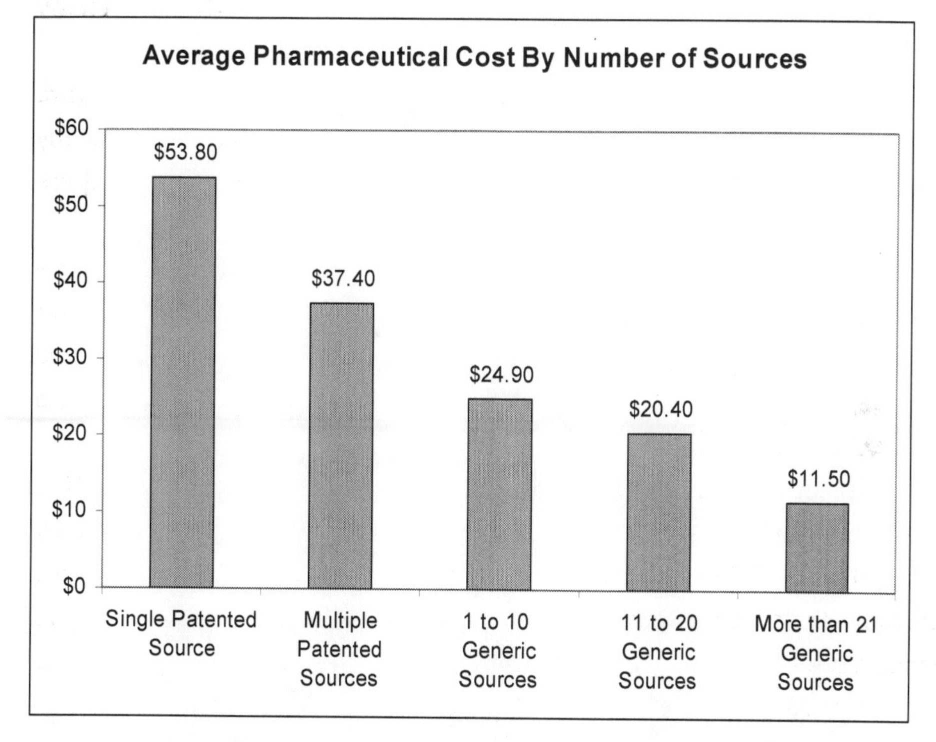

Figure 1 - As more sources manufacture a drug to meet consumer demand, prices drop.

In a way what's happening here is a shift in control from corporations to consumers. While patent protection lasts, the inventor has complete control over how much of a drug is made and how much to charge for it. Once patent protection is gone, manufacturers must scramble to outdo each other in bidding for the customer's dollar.

How low can that competition drive the price of a drug? Again, the greater the demand, the lower the price. Penicillin was one of the greatest breakthrough drugs of the past century. Initially, it was also extremely expensive. For decades after its discovery, the drug was effectively priceless – there was no method to produce it in bulk. With the demand for antibiotics of World War II, large scale synthesis methods were developed. By the end of the war, the price of penicillin had dropped to around twelve dollars per gram. Over the next thirty five years it continued to drop, to a current cost of less than 2 cents per gram, or less than one five hundredth its price in 1945.

There's no reason to believe that new enhancement drugs will be substantially more expensive to produce than penicillin if manufactured in equivalent volume. Gene therapies will be a bit more expensive at first, solely because they're manufactured through new and different techniques. But fundamentally they're information products in much the same way that drugs are. The cost of manufacturing is almost insignificant when compared to the cost of research, development, and testing.

What does this tell us? It means that the first effective performance enhancing drugs may well be expensive at first. A company that had a drug that doubled your learning rate could charge an incredible amount for it. A company that had a drug that cut your rate of aging in half could charge even more.

Yet once the patents on truly effective performance enhancing drugs have expired their prices will drop in proportion to the number of people who want the drug. The most sought-after enhancements will be the cheapest. A drug that significantly slows aging, for example, might be in as much demand as penicillin, and therefore just as cheap - pennies a day. If so, such a drug would be within the buying power of most of humanity.

Of course, not all enhancements take the form of drugs or injections of gene vectors. Some enhancement techniques will require considerable time or equipment. Yet over time these sorts of procedures drop in cost as well. Consider laser eye surgery. Initially there were only a few LASIK clinics in the world. Those clinics often had no competitors in the same city. In the first year that LASIK was available, only twenty thousand procedures were performed. The average cost per eye was around $4,000. But as more ophthalmologists entered the business of providing LASIK, the cost dropped and the number of procedures soared. By 2000, more than 1.5 million eyes were being sculpted by laser each year, and the cost was as low as $500 per eye.

$500 still isn't in the budget of most people on earth. Yet the cost of LASIK has dropped as much as it has while the patents that cover it are still held by just two companies: VISX and Nidek. When the first of those patents expires in 2007, the cost is set to drop even further, bringing it into the hands of more people.

This suggests that labor-intensive enhancements are also controlled by a kind of inverted supply and demand rule. Highly in demand services will attract providers. The competition between those providers will drive prices down, shifting control to consumers.

Not all medical procedures follow this pattern. Most surgeries are paid for by insurance. Since insurance companies set the price they're willing to pay, and the patients choosing the doctor seldom have much awareness of the overall price, there's not the same competition to push costs down. However, for purely *elective* procedures where the patient pays the price, the rule applies: Greater demand drives prices up in the short term, but down in the long term.

The other reason to believe that legal enhancement technologies could benefit the poor almost as much as the rich is the law of diminishing returns. The better off you already are, the more you have to spend to increase your well being at all.

Consider automobiles. In most places in the United States, a few thousand dollars will purchase a fairly safe, comfortable vehicle for getting from point A to point B. Alternately, many tens of thousands of dollars will buy you a new luxury vehicle. A luxury car can cost ten times what the most affordable new economy car would cost. Yet the luxury car isn't ten times as fast, or ten times as fuel efficient, or ten times as safe. The extra money spent on it buys only incremental advantages. In terms of basic mobility, the inexpensive difference between no car and a cheap car is larger than the very expensive step from a cheap car to a BMW.

This law of diminishing returns applies to other aspects of life as well. One good measure of human well being is life expectancy. To have a high life expectancy one must have access to food and water, adequate shelter from the environment, easily available health care, vaccination against major diseases, protection against violence and warfare, an environment free of pollution, crowding, and stress, and so on. What's more, life expectancy correlates well with education, with political freedom, with equality within a country, and other positive social indicators. Overall it may be the best single proxy for the various components of human quality of life.

Like the quality of a car, life expectancy goes up as you spend more money, but more and more slowly as spending increases. The graph below shows the life expectancy and per-capita Gross Domestic Product (GDP) of 175 countries. Higher per-capita GDP moves a country towards the right. Higher life expectancy moves a country towards the top. The trend line that cuts through the dots initially shoots up towards longer life and then flattens out. Greater national wealth strongly affects life expectancy at the low end. But after per capita GDP of about $3,000, more money has only a small impact on life expectancy.

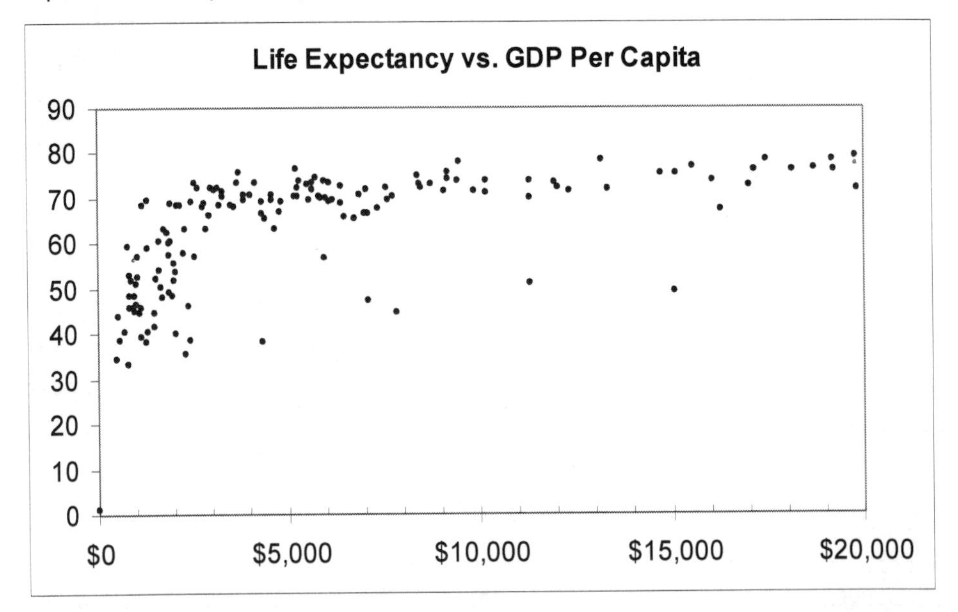

Figure 2 - Life expectancy rises quickly with increases in per-capita GDP, then levels off.

The same pattern is generally true if we look at life expectancy vs. per capita health spending instead of per capita GDP. Going from no health care spending to $300 per person per year – less than a dollar a day - makes a substantial difference in life expectancy. Going from $300 to $300 makes almost no difference.

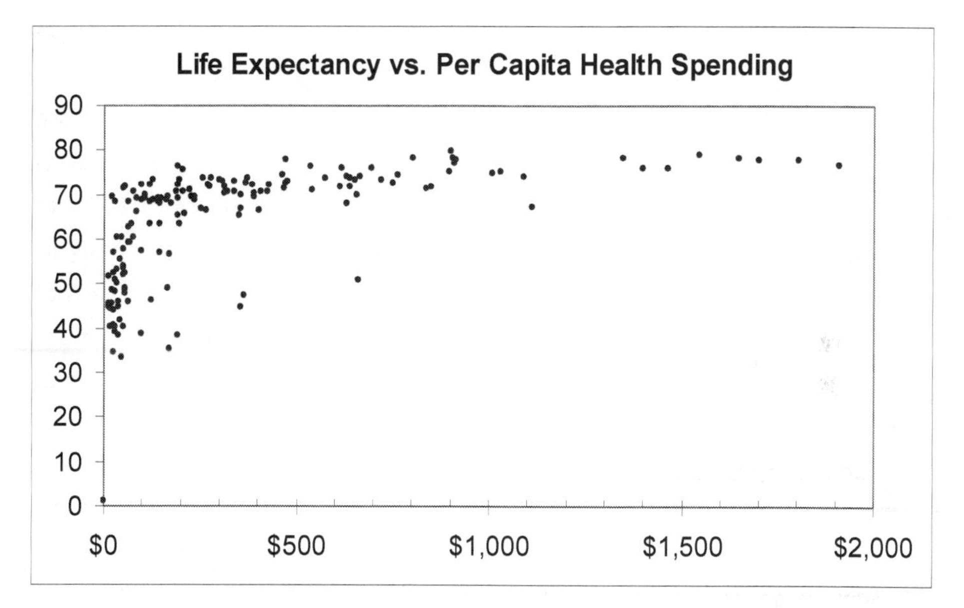

Figure 3 - Life expectancy rises quickly with increases in per-capita health spending, then levels off.

In fact, the largest increase in life expectancy occurs by the time a country reaches around $100 per person year in health spending. Additional spending increases life expectancy, but not as much as that first hundred dollars. Basically, the poor are getting more improvement in their health per dollar than the rich. Alternately, you can think of it in terms of the price of years of life expectancy. The first few years come cheap. The later years get progressively more and more expensive.

If enhancement technologies follow the same pattern, then what we'll see is not so much a gap between rich and poor as a time lag. Every new enhancement technology will initially be fairly expensive, and thus available only to the rich. In time, the technology will have competitors, or will go off patent, and its price will drop markedly. For any given type of enhancement there will always be newer, more expensive techniques that work slightly better than the older, cheaper techniques. At any given time the rich will be able to afford the latest techniques while the poor will not. But if the relationship between life expectancy and spending is any guide, the latest and greatest enhancements may cost anywhere from 10 to 100 times as much as older techniques while offering only a slight improvement. Both rich and poor will benefit, with the rich paying dearly to benefit first.

ON THE OTHER HAND, if governments restrict access to enhancements, by attempting to ban them, for example, prices rise. Consider the cost of tobacco and marijuana. Both are easy to grow and widely used in the United States. There is nothing about the plants that should suggest a radical difference in price. Yet while tobacco sells for a few dollars an ounce, Marijuana costs around $300 an ounce, or *one hundred times* more.

Higher prices of illegal drugs may or may not be a bad thing. Higher prices of enhancements, however, would exacerbate the gap between rich and poor. The more effective a ban, the higher the price of a black market enhancement will be. As prices climb, the population able to benefit from an enhancement shifts towards the rich.

In the extreme case, in which a country is able to perfectly enforce a ban on an enhancement, the rich will simply go overseas. Attitudes towards enhancement are quite different in different parts of the world. While the US and other western nations may ban genetic engineering or other enhancement techniques in the coming years, those techniques are likely to remain legal in Asia and other parts of the world.

The richest westerners will always have the ability to go abroad for their enhancements. A wealthy couple can choose to fly to China or some other location where an enhancement procedure is legal, stay for the necessary days, weeks, or months, and then return. If the procedure can be imported in the form of a drug or gene therapy, they're more likely to be able to pay the high cost of smuggling. A poor or middle class couple doesn't have the same options. Prohibition would effectively put enhancements farther out of the reach of the poor. If equality of access is a goal, it makes more sense to encourage the development and use of enhancement techniques, to spread it to as many people as possible, rather than restrict it.

Regulating Enhancement

An important step in spreading the power to alter our minds and bodies to more individuals is to reform the way that enhancement technologies are regulated. A drug or gene therapy that enhanced memory or learning or slowed the aging process wouldn't necessarily be illegal today, but it would go through a regulatory process that would make it harder for doctors and consumers to get high quality information about the safety and effectiveness of the technique.

In the U.S., drugs, gene therapies, and medical devices are regulated by the Food and Drug Administration (FDA). Gene therapies are regulated essentially as drugs. Medical devices are regulated fairly similarly. To be approved for human use in the United States, any drug or gene therapy must go through three stages of development.

First, pre-clinical research using chemical analysis, cell cultures, and animals must indicate that the drug is effective in treating a disease and has a high likelihood of safety in humans. If this pre-clinicial stage is successful, a researcher or pharmaceutical company can submit an Investigational New Drug application, or IND. If the FDA determines that animal and lab experiments have provided enough evidence of safety and effectiveness, the agency will approve the IND, and the drug moves on to human trials.

Human trials occur in three phases, Phase I through Phase III. Phase I studies are about safety. They use low doses in a small number of patients to look for any potential side effects. Phase II studies focus on effectiveness, looking at the impact of the drug on a larger population, usually a few hundred people, to determine if the drug is actually effective at treating the targeted disease or condition, and continuing to monitor for safety problems. Phase III studies look at effectiveness and safety in a yet larger population –

this time several thousand people - over a period of many months to a few years.

Since the clinical phase involves at least three separate sets of human experiments, with time for analysis of the data between each one, it takes a minimum of several years, and at times as long as a decade. At the end of this period, if the drug is effective in treating the disease and safe in large populations for extended periods of time, the sponsor can submit a New Drug Application or NDA. If the FDA approves the NDA, the drug can be marketed and sold for the specific condition it was approved for.

Once a drug is on the market, physicians are allowed to prescribe it for "off-label" use so long as that use falls within a nebulous definition of "the practice of medicine". As a practical matter, the FDA scarcely regulates physicians, so for most drugs doctors are free to use their best judgment in prescribing it for related conditions. For example, a doctor would be effectively free to prescribe an Alzheimer's drug that also improved the memory of healthy young people to a college student cramming for an exam. Whether or not the doctor would do this is up to her. The major exception to this rule is any drug (typically those with recreational uses) defined as a Controlled Substance. Prescriptions of those drugs are monitored by the Drug Enforcement Administration.

While physicians are essentially free to prescribe drugs for off-label uses, pharmaceutical companies are much more restricted in what they can advertise. The FDA prohibits drug manufacturers from advertising any off-label use to either physicians or patients without explicit permission of the agency, except in special cases. And since drugs can only be approved for use against a disease, the "on-label" use will never mention any enhancement purpose, even if sound scientific research backs up the effectiveness of the drug as an enhancement.

Interestingly, the FDA's rules are reversed when dealing with dietary supplements. Labels and advertisements for dietary supplements cannot make any medical claims. A vitamin bottle, for instance, can't carry the claim that it lowers the risk of cancer. However, dietary supplement packages can make virtually any other kind of claim, without any scientific basis. So an essentially untested herbal product can bear a label which claims "Improves Your Memory!" while a well tested drug which has been scientifically shown to improve your memory can not.

The FDA's lack of recognition of enhancement uses of drugs has two other effects. First, it reduces the incentive of a manufacturer to run the experiments that would determine whether a drug or gene therapy actually enhances those who use it. Since a pharmaceutical company can't market a product on this basis, there's a lower return on investment from any such studies.

Second, the safety trials required before approval of a drug or gene therapy typically focus on people with the disease or condition being targeted. As a result, safety in people who *don't* have the disease isn't tested. A gene therapy technique for muscular dystrophy won't be tested on athletes seeking to increase strength, and thus the safety of the product for athletic use won't be known.

The net result of the current regulatory scheme is that consumers and doctors are denied useful information about the safety and effectiveness of potential enhancements. The regulatory model could be improved quite simply by doing two things.

First, the FDA and its analogous bodies in other countries should evaluate the effectiveness of drugs for enhancement claims as well as disease-treating claims. In most cases the drugs that manufacturers wish to make enhancement claims for are also under review for medical uses. Testing and approval of a drug for an enhancement claim would be virtually identical to the process for drugs today, except that it would involve safety testing on a wider population, and would also require verification that the drug actually enhances mental or physical function in the way its sponsors claim.

This small change to the FDA process would bring three beneficial effects. Doctors and physicians would now have reliable information about which drugs actually enhanced human function. They would also have reason to doubt the effectiveness of any drug that *had not* been approved by the FDA for an enhancement use. And we'd know more about whether drugs already being consumed as enhancements are safe when used in that way. What's more, opening up the enhancement uses of a drug would increase total demand, driving down prices for both enhancement and therapeutic uses.

A second, slightly less important change would be to restrict the enhancement claims that dietary supplements can make without research to back them up. This would help consumers distinguish between products for which there's scientific evidence of enhancement effects and those for which there's not. Practically speaking, this change is unlikely to happen in the near future. The FDA proposed a similar policy after the passage of the Dietary Supplement Health and Education Act in 1994, but lobbying pressure from the dietary supplement industry forced the agency to weaken the rules to only restricting "medical" claims.

Independent of restrictions on dietary supplement claims, scientific testing of the safety and effectiveness of enhancement drugs and gene therapies would educate and protect consumers, and offer them a scientifically validated alternative to less regulated products whose claims are rarely backed by evidence.

IF GOVERNMENTS REFORM their regulatory processes to provide doctors and consumers access to the best possible information about enhancement techniques, then market forces will probably suffice to bring the cost of enhancement technologies down to the level where the bulk of society can afford them. Governments, in general, are well served by allowing the market to do its job, thus leveraging the intelligence of millions of individuals who can choose the best product for the price, and reward the companies that bring them to market.

At times, though, the market does fail, or fail to bring prices down quickly enough to meet a pressing need. If this occurs with enhancement technologies, then governments should step in to reduce the cost.

Inequality in access to enhancement technologies brings the risk of stratification of the rich from the rest of the population. Some enhancements

– like those to learning ability or memory – will increase earning ability. If the rich are able to buy these enhancements while the poor cannot afford them, then those who start out rich may pull further away from the poor, using their money to buy better enhancements, which earn them more money, which allow them to buy even better enhancements, and so on. For the rich this would be a virtuous cycle of gains begetting more gains. For the poor it would be a vicious cycle, with lack of enhancements preventing access to the best jobs, thus robbing them of the money they need to buy enhancements.

The recent history of technology has *not* followed this pattern. In real terms, the poor of the world are closing the gap with the rich – in life expectancy, in education, in access to technology, and so on. Like other technologies, biotech enhancements are likely to rapidly drop in price. The only reason to fear such a stratifying effect of biotech enhancements is their potential for rapid and dramatic effects. If biotech and neurotech enhancements increase earning power *faster* than they drop in price, they could lead to runaway stratification. In this case, governments should act.

Governments should also be prepared to step in as a matter of opportunity rather than risk. As previous chapters have shown, enhancements that boost mental function can contribute to economic growth, and enhancements that slow aging or reduce disease can save society large sums of money that would otherwise be spent on healthcare. In the United States alone, every 1 percent increase in productivity adds $100 billion to the economy, and every 1 percent reduction in health care costs saves $28 billion.

Given these impacts, biotech and neurotech enhancements can be seen as investments in valuable human capital. In enhancements are cheaply and widely available, then individuals and families will decide where and how much to invest them. If for whatever reason the market fails to reduce costs to the point where most people can afford to purchase biotech augmentations of their minds or bodies, then governments have a tremendous amount to gain by investing their own funds in spreading access to mental and health enhancements to the public.

Government spending in these areas has precedents. England, Australia, Japan, and other countries already subsidize the cost of in-vitro fertilization, with the effect that in Australia the procedure is four times more common than in the United States.

Governments also invest in enhancement in other ways. In the United States, the federal, state, and local governments spend a combined $6,911 per student for elementary schooling in 2000. That totals nearly $382 billion per year. Over the lifetime of a student, that comes out to nearly $83,000 spent on public education, not considering the cost of college.

By contrast, most enhancement techniques would be rather cheap. The best comparison for gene therapy costs, for instance, might be vaccinations. The per capita costs of influenza vaccines is around $15 in the United States today. In comparison to the money society already spends, biotech enhancements could provide a dramatic return on investment.

SOCIETIES ARE FORMED to secure the rights of individuals. In the west we embrace the freedom to make ourselves who we want to be. We work to increase equality - not a stifling sameness from person to person, but equality of *opportunity* - so that as many individuals as possible might possess the power to steer and improve their own lives. History shows us that when individuals are given the power to make such choices, it serves to benefit society as a whole. It's also shown us that human beings are the ultimate resources, the most precious capital a society has. It makes good sense, both on grounds of principle and pragmatics, to invest in this resource. Rather than trying to restrict freedoms, governments ought to focus on empowering individuals and families, and reaping the benefits for all humanity.

Chapter 4 – Methuselah's Genes

Benjamin Franklin famously observed that nothing is certain in life except death and taxes. That hasn't stopped people from trying to avoid either, and while taxes have remained with us, we've made progress of a sort against death. A child born in 2000 will live, on average, to the age of 66, roughly twice as long as a child born in 1900, and three times as long as a someone born a millennia ago.

Humans live longer, on average, than ever before. Yet until recently this progress has been made not by combating aging, but by protecting us against the things that kill us while we're young. Medicine has done very little to lengthen the lives of those who make it to old age. In 1900 a man who'd managed to survive to age 70 in the United Sates could expect to live to age 79. Today a 70 year old American man will live on average to age 82. That's a gain of just three years of life expectancy at that age over the last century – a trifle compared to the gains for newborns.

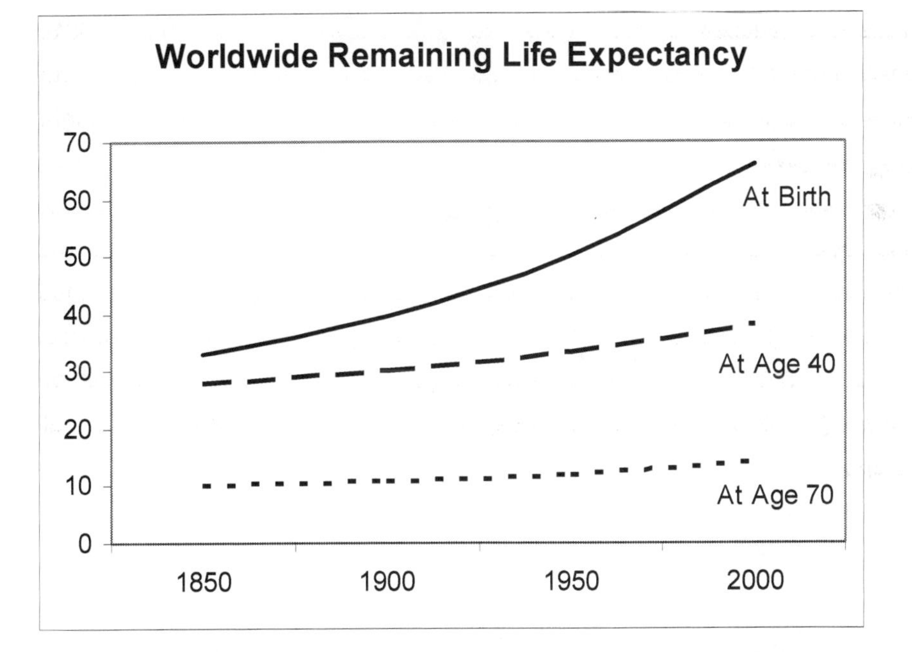

Figure 4 - Life Expectancy at birth has gone up sharply over the last 150 years. Life expectancy at later ages has risen more slowly.

We've made such small progress in increasing life expectancy at more advanced ages because medicine has focused primarily on disease rather than the aging process itself. Antibiotics have lengthened the average human life, but haven't changed the fact that our immune systems get weaker as we age. Heart surgery can repair acute damage, but can't restore the heart to youthful health. Nothing we've done so far has addressed the fundamental issue: The older we get, the less capable and more vulnerable to health problems we become. If we are living longer, it's because more of us are living to be *old*, when what we really want is to *stay young*.

The vulnerability that comes with advanced age is captured in something called the Gompertz Law. Benjamin Gompertz was a self-educated Londoner living around the turn of the 19th century. His law states that in any species, the risk of death rises exponentially with age. Applied to humans, this says that our risk of dying from any cause roughly doubles every 8 years. In essence, time wears down our minds and bodies. The diseases that kill us in old age - heart attacks and cancer, for example - are largely symptoms of the aging of our bodies. Every time medicine cures a disease and lengthens life, humans run into the wall of another disease. Alzheimer's disease is an example - as people in developed countries live longer, more and more suffer from this disease.

Demographer Jay Olshansky of the University of Chicago says as much, "If, indeed, we were to find a cure for cancer, life expectancy at birth would rise by about three and a half years. ... And if we eliminated all cardiovascular diseases, diabetes and all forms of cancer combined, life expectancy at birth in humans would rise up to about 90." That's a roughly ten year rise in life expectancy, a great advance, but less than a third of the gain humanity made in the twentieth century.

In Swift's *Gulliver's Travels* there's a scene that puts our progress in lengthening human life in stark relief. It's the story of a race called the Luggnaggians, and the immortals among them called Struldbruggs. Most Luggnaggians are ordinary, mortal creatures who die when their time has come. Struldbruggs are rare Luggnaggians who nature has gifted (or cursed) with immortality. The Struldbruggs age but do not die. As time passes they grow weaker, frailer, and more miserable.

This notion of a frail, ailment-ridden old age jibes with our fears for our own last years - that we'll spend them hooked up to life-preserving machines, suffering through dozens of bodily ills, and mentally addled enough to require constant supervision. We all want to live, but not this way.

Critics of techniques to lengthen the human lifespan often paint a Struldbruggian picture of our future. In these scenarios, a booming elderly population is kept alive only through incredibly expensive medical techniques paid for by the young. And in fact, there's some justification for such scenarios based on the current trends in health care and life expectancy. As the population of the developed world ages, pension programs will come under pressure, medical costs will soar, and the number of young able-bodied and able-minded people available to pay for it all will shrink in comparison.

This underscores the core point - if we want to live longer, we must begin addressing not only the symptoms of old age, but the phenomenon of aging itself.

Actually slowing aging - extending youth - would pay substantial dividends. It would reduce the incidence of age-related diseases that account for roughly half of all deaths today. It would reduce the cost of healthcare across the world by hundreds of billions of dollars, as the young generally consume less medical care than the old. It would keep people mentally and physically fit longer, enabling them to pay their own way through society, rather than draw on the income of younger generations. It would, in short, address the very problems that longer lives are causing.

A pill or injection that prolonged youth would also prove immensely popular. In in 2002, consumers in the US alone spent an estimated $3.6 Billion on supposed 'anti-aging' products ranging from anti-oxidants to growth hormone injections, despite the fact that not one of them has been proven to slow or reverse aging. Some of these products are sold by true believers. Just as often, they come from unscrupulous vendors who believe only in the money they can make selling so-called cures for aging to scientifically naïve customers.

The market for scientifically unproven 'anti-aging' products has so disturbed real scientists in the field of gerontology that fifty-one of them signed a joint 2002 "Position Statement on Human Aging". Their paper stated unambiguously that today, for human beings "there is no such thing as an antiaging intervention."

The authors of the paper harshly critiqued existing 'anti-aging' products on the market. Their goal was to educate a public that, in their eyes, is being duped by blatantly false advertising. Yet after debunking today's products, the authors ended on a more optimistic note, stating that "our rapidly expanding scientific knowledge holds the promise that means may eventually be discovered to slow the rate of aging." In the pages that follow we'll talk about some of those promising discoveries.

The Future of Mortality

"In the past few years a number of experiments have completely blown away the theory that aging could not be slowed," says biologist Gordon Lithgow. Lithgow should know. He's one of a vanguard of scientists who've found techniques that have actually slowed aging in animals and which, one day, may do the same in humans.

Andrzej Bartke, one of the signatories of the 2002 position paper, also knows first hand the potential of science to slow aging and extend life. On January 8th of 2003, a genetically modified mouse in his lab at Southern Illinois University died just a week short of its 5th birthday. Average lab mice die at about half that age. The very longest lived lab mice live for three years. Bartke's genetically enhanced mouse lived two thirds longer than the maximum lifespan of his species – as if a human had lived to be 200. And it lived that time in good health, free of the diseases and fragility that plague most old age.

It was the work of geneticist Tom Johnson at the University of Colorado in the late 80s that broke open the field of age interventions. Johnson started working on the genetics of aging in 1979. At the time he believed, like most other researchers, that aging would be so complex that it would require changes to hundreds or thousands of separate genes to extend the life span of an organism. Then in 1983, his colleague Mike Klass found several mutant strains of the tiny nematode worm that had exceptionally long lifespans - up to twice as long as the norm for their species.

After Klass left academia for industry, Johnson took over his work. Johnson initially assumed that to have a lifespan twice as long as normal, the worms must have had mutations to a large number of genes. He gave the project to David Friedman, then an undergrad student working under Johnson. What Friedman found was that, contrary to what anyone suspected, the long lived nematodes had a mutation of just a single gene (later named *age-1*) out of the 19,000 in the creature's genome. In papers in 1988 and 1990, published in *Genetics* and the prestigious journal *Science*, Johnson and Friedman laid out their amazing findings - a change to a single gene could double the lifespan of the nematode worm. Later research would go on to show that mutations of two genes combined could even triple lifespan.

University of Idaho evolutionary biologist Steven Austad, another leading researcher in the field of aging, underscores just how tiny a genetic change this is, "Worms have 100 million nucleotides in their genome approximately. If you change one of those, one of the letters, one of the 100 million letters of the genetic alphabet, you get a doubling of life span. If you change two of the letters, if you change one letter in one gene, one letter in another gene, you'll get a tripling of life span."

While Johnson's research opened the way for much of the field of life extension, it wasn't well received at first. "My initial findings," he says, "were pretty much ignored by most people. ... I really felt ostracized. I remember very much a comment by a colleague once to one of my students to the effect that "you are so smart what are you doing working with that charlatan? Some day the world is going to find out that he has been faking data."

In the end, the scientific method won out over dogma. Five years after Johnson's publication of his results, Cynthia Kenyon at the University of California in San Francisco – already a well known researcher - discovered that another gene called *daf-2* could also lengthen the life of these worms. With two different labs showing similar results with different genes, it was harder to claim that the life extension was an error or a hoax. Prominent scientists started to take notice of this idea that small genetic changes could extend life, and set about searching for such genes themselves. The field of genetically extending life was born.

Since then, scientists have discovered dozens of genes that can each increase lifespan in a wide variety of creatures. Some of them extend life by just a few percent, while others can more than double life span. Scientists now know of more than a hundred genes that can slow the aging process in animals like nematodes, fruit flies, worms, and mice. More are being discovered every day. These genes that researchers are altering aren't just present in animals – they're present in us humans. That suggests that the same genetic changes in men and women might slow human aging. A host of researchers in both academia and private industry are working on ways to use this genetic knowledge to design new life-extending drugs – drugs that might lengthen our lives by decades or more.

Life extension genes fall into a handful of different families. Genes of the earliest discovered family have variants which reduce the body's sensitivity to insulin or a related chemical called insulin-like growth factor. A second and more recently discovered family of genes works by protecting cells against the ravages of free radicals – highly reactive chemicals that bounce around causing damage to whatever they touch.

Age-1, the gene that Johnson discovered, is in the insulin-regulating group. Today this is the best studied family of life extension genes. Genes of this sort show up in almost every creature alive. Scientists have lengthened the lives of nematodes, fruit flies, and mice by making changes to these genes.

Insulin and its cousin insulin-like growth factor 1 (or IGF-1) are messengers. The body uses them to send signals from one part to another.

For example, when a muscle cell is contacted by a molecule of insulin, the muscle cell responds by sucking up more sugar from the bloodstream and storing it away as a chemical called glycogen for future use as fuel. *Age-1* and its related genes are involved in making the *receptors* on each cell that pick up these chemical messages. Johnson, Kenyon, and other researchers have extended the lives of nematodes by reducing the function of *age-1* and related genes, making the animal's cells less sensitive to insulin.

These genetic changes have also made the worms more resistant to stresses of all sorts. Between 1993 and 1995, Biologists Simson Melov and Gordon Lithgow at the Buck Institute for Age Research ran a series of stress experiments on *age-1* and *daf-2* nematodes. They put the worms in high temperatures that would kill normal creatures. They also exposed them to toxins that they shouldn't be able to survive. In both cases, they found that the worms that had been genetically modified for longer life were also more resistant to the external stress. This made the genes even more attractive as targets for life-extending drugs in humans. After all, no one wants to be old and frail, but a single treatment that could keep you young and also protect you against toxins and other stresses sounds pretty appealing.

Exactly how and why mutations of the insulin pathway slow aging remains something of a mystery. It may be related to the protection against stress that Melov and Lithgow discovered, though. In a review published in *Nature*, researchers Toren Finkel and Nikki Holbrook at the National Institutes of Health pointed out that virtually every age-slowing genetic change discovered to date confers added resistance to stresses like heat and toxins that could normally damage DNA. By contrast, virtually every genetic mutation that accelerates aging also makes creatures more vulnerable to heat, ultraviolet light, toxins, or other environmental stresses that could damage their genes. If Finkel and Holbrook's hypothesis is correct, the insulin-like pathway in the body plays a role in protecting our cells against genetic damage, and the changes to this pathway that slow aging increase its protective function.

Some life extended creatures do suffer from other side effect, though. Kenyon's *daf-2* mutants experienced reduced fertility and other problems that seemed tied to their childhood development. By taking genetically normal nematodes and reducing the function of *daf-2* only in adulthood, Kenyon was able to extend their lives while avoiding these problems. Making the change at adulthood brought all the benefits but none of the side-effects.

A more important question than how these genes work in nematodes is how they work in mice. Mice are mammals, more similar in their biology to humans than insects or worms, so we'll learn far more about the prospects of extending human life by studying them than nematodes or fruit flies. The quest to lengthen the lives of mice by reducing their sensitivity to insulin or IGF-1 has been tricky. In 2000 and 2001 a number of studies showed that it could be done, but at a price. Four different labs produced "dwarf" strains of mice that either lacked or could not respond to growth hormone. One of the ways that growth hormone affects the body is through the release of insulin like growth factor. So by blocking the effects of growth hormone, the researchers indirectly blocked the effects of IGF. The modified mouse strains live anywhere from 30% to 60% longer than normal mice. Unfortunately, the mice all experienced serious side effects. Growth hormone is essential in development. Without it, the mice ended up physically stunted - around half the size of other mice. Two of the strains are almost completely infertile.

While the side effects of reducing growth hormone function were a blow to the study of the insulin and IGF-1 anti-aging techniques in mice, they didn't kill the idea of modifying IGF-1 related genes to extend life. The first four studies all made larger changes to the mouse's genes than anyone wanted. No one wanted to disable growth hormone for its own sake - reducing growth hormone was just a handy way to reduce the function of IGF-1. So, was it possible to make a more precise change to the mice? Could researchers reduce the function of the IGF-1 receptor without knocking out some other important part of the mouse's biology? If so, would that very targeted change extend life? And would it have major side effects?

Two partial answers came around the end of 2002 and beginning of 2003. In late 2002, French geneticist Martin Holzenberger reported that his team at the National Institute of Health and Medicine in Paris had created mice that had only one copy of the gene for the IGF-1 receptor, instead of the normal two copies. Up until then no one had succeeded in doing anything so specific. Holzenberger's mice live longer than his controls - 16% longer in the case of the males and 33% longer in the case of the females. They're also full size, ate as much as normal mice, and seemed just as active. And they're more resistant to stress - they survive exposure to a toxin called paraquat hours longer than did the control mice.

Holzenberger's study had a few problems, though. The strain of mice he started with, before his genetic modifications, wasn't very long lived, especially in the rather crowded conditions in which the mice were housed in this experiment. As a result, his life-extended mutant mice, which lived longer than the strain he started with, still died earlier on average than other strains of mice kept in better environments. Still, Holzenberger clearly showed that by modifying the IGF-1 gene he could lengthen the life of one strain under one set of conditions. Combined with the evidence in nematodes and other strains of mice, this suggests that IGF-1 will lengthen the lives of mice living in better conditions as well.

A related study showed that mice that lacked an insulin receptor in their fat cells lived longer. Researchers at Harvard increased the average live span of mice by about 18% this way. The mice looked and acted normally. They had less body fat but were otherwise normal size. And they ate about one and a half times as much as other mice of the same size but remained thin and healthy, and still outlived the genetically unmodified mice. Double bacon cheeseburger fans take note.

All told changes to the family of genes that controls the insulin and insulin-like receptors have extended life in every species they've been tried in. Researchers have even found that they lengthens life in yeast. Now what everyone wants to know is, will such changes work in humans?

There are two reasons to believe that these genes could be used to increase human lifespan as well. First, *age-1*, *daf-2*, and other life-extending genes that we haven't talked about are very similar to genes that humans carry for the insulin and IGF-1 receptors. So the molecular switches we've flipped in nematodes, flies, and mice exist in our genetic code as well.

Second, it's telling that IGF-1 related changes have increased the lifespan of mice, fruit flies, nematodes, and yeast - incredibly different species. Mice are relatively close relatives of ours, having diverged from us on the evolutionary tree around 75 million years ago. Nematodes, fruit flies, and yeast, on the other hand, diverged from us hundreds of millions of years ago, or more than a billion in the case of yeast. Yet, the same life extension techniques seem to work on all of them.

This is a profound discovery. The same techniques extend the lives of warm blooded creatures, cold blooded creatures, and creatures without blood at all. They work on species that reproduce by budding, species like nematodes where every member is a hermaphrodite, species that reproduce by laying eggs, and species that have sex and nurse their embryos in the womb, like us. The same techniques lengthen the lives of nematodes with less than a thousand cells and mice with more than a trillion cells. They slow the aging of creatures we diverged from as long ago as a billion years and as recently as 75 million years ago. They prolong the youth of animals with brains with billions of neurons and creatures with no brains at all. Given how different these species are, the IGF-1 gene changes must be affecting aging at a very fundamental level.

Cynthia Kenyon, for one, believes that this broad spectrum of species provides the crucial evidence that the IGF-1 gene can be used to extend human life as well. Speaking about Holzenberger's work in mice. "It's wonderful," she says, "It shows quite clearly that [the insulin-related longevity pathway] is conserved in mammals, and that's really powerful. If it's true in worms and flies and mice, the chance that it's not true in humans is small."

Kenyon is confident enough to bet her time and her career on it. In 1999, she co-founded Elixir Pharmaceuticals with prominent aging researcher Lenny Guarente. Years earlier, Guarente had discovered that a gene called Sir2 could lengthen life in yeasts and nematodes. Sir2 seems to block the activity of harmful genes inside the body. Humans and other mammals carry descendents of these genes called sirtuins, and researchers are currently looking at lengthening the lives of mice by modifying these genes.

Between them, Guarente and Kenyon have more than 250 scientific publications to their names. Along with Tom Johnson, they're among the most respected researchers in the field of genetically modifying animals to increase their lifespans.

In late 2002, Elixir merged with Centagenetix, a company looking at the genes of exceptionally long-lived humans to find drug targets to cure age-related diseases. They've raised between $20 and $30 million to fund their research. Elixir's strategy for finding drugs that lengthen life is simple. Since making cells less sensitive to IGF-1 extends life in so many animals, they'll look for drugs that can do the same thing artificially.

The IGF-1 receptor, like all other receptors, is a bit like a three dimensional puzzle piece. A molecule of IGF-1 fits snugly into the pocket of the IGF-1 receptor. This tells the receptor that a signal has been received, and the receptor in turn notifies the rest of the cell. One way to reduce the sensitivity of the receptor is to find a molecule that makes it harder for an IGF-1 molecule to fit snugly into the right place. Imagine inserting an odd-shaped piece of cardboard inside the pocket that is the IGF-1 receptor. The foreign piece of cardboard would make it harder for the other puzzle piece (the molecule of IGF-1) to fit in the right place. A molecule that does this is called an *antagonist* because it works against the function of the receptor.

Elixir is in the process of searching for compounds that might work this way. After they find a good target they'll test it in lab animals. If the drug extends life, and if it's safe, then they'll slowly and carefully move to tests on humans. The whole process, if it succeeds, will take 10 years or more. Of course, the FDA and other regulatory agencies only approve drugs to cure diseases, and they don't consider "aging" a disease. So the most likely strategy for Elixir is to show that their drug helps stave off some symptom of aging, like cancer or diabetes or heart disease. While such a drug it wouldn't be advertised as slowing aging, some people would probably seek it out for that reason. History shows that they'd find sympathetic doctors to prescribe it – just as motivated patients have a fairly easy time acquiring Viagra, antidepressants, or other prescription drugs from their physicians for off label use.

To prevent patient harm, regulatory bodies will need to insist on tests of these drugs designed to show how they affect perfectly healthy people not yet suffering from cancer, diabetes, or heart disease. For now, the scientific and business communities aren't worrying about the regulatory issues. If drugs can be found to slow human aging, there will undoubtedly be a huge market for them. Even if Elixir isn't the company to find such a drug, the search is on. Researchers of aging around the world now know that variants of genes related to IGF-1 extend life in a variety of creatures. Other labs have identified additional genes in this family, and pharmaceutical companies have their eyes on the field.

Free Radicals

The other large and growing family of genes that extend life are those that protect the cell against free radicals. Free radicals are dangerous molecules inside every cell of your body. They're electrically charged molecules that can easily chemically react with other parts of the cell. In particular, they have a high affinity for electrons – they will frequently bounce into another molecule and steal an electron from it. When a free radical bounces against a strand of DNA and strips off an electron, it can cause a mutation. Over time these mutations build up, causing your cells to malfunction in subtle and then progressively less subtle ways.

Your cells do an amazing job repairing the damage free radicals cause, but occasionally something gets past their self-repair machinery. Much of the damage from free radicals accumulates in the poweplants inside your cells called *mitochondria*. As mitochondria produce the energy for the rest of your cell, they create a large number of free radicals as byproducts. Those free radicals often damage the small amount of DNA in the mitochondria themselves. As damage builds up in these cellular powerhouses, they're less able to produce energy for the rest of the cell. Over time these problems build up. Cells with damaged mitochondria can't produce as much energy, so they just can't work as hard. Muscles become weaker. Brain cells become slower. Some cells just stop working completely.

Researchers have uncovered the effects of free radicals and mitochondrial DNA damage in experiment after experiment. Different species of animals produce different amounts of free radicals. Longer lived species produce fewer free radicals, and shorter lived species produce more free radicals. Researchers have also created tests to measure the amount of damage that free radicals do. They've found that in creatures with short lifespans there are lots of free-floating proteins and other molecules that have been damaged by free radicals. In longer-lived species the concentration of free-radical damaged molecules is much lower.

Scientists studying nematodes have also found that older nematodes have more damaged mitochondrial DNA than younger nematodes. And life extension genes like *age-1* slow the rate of this damage. In fact, most of the life-extension genes we've talked about seem to slow the rate of free radical damage and make creatures more resistant to toxins that trigger free radicals.

Most importantly, genes that fight the effects of free radicals can also slow aging. Antioxidants are chemicals that stabilize free radicals and make them harmless. Vitamin C is a popular antioxidant. Unfortunately, experiments in feeding animals vitamins haven't lengthened their lives. But experiments that have increased the body's supply of its own antioxidants have done wonders. Inside the body, two of the most important antioxidants are chemicals called superoxide dismutase (SOD) and catalase.

In 2002 scientists created two strains of fruit flies that produce more SOD. These strains live about 40% longer than normal fruit flies. Another genetically modified strain of fruit flies creates extra methionine sulfoxide reductase A (another important antioxidant in the body) and also lives substantially longer than normal flies. A fourth strain of fruit flies carries a gene whimsically named *methuselah* that gives it extra protection against free radicals. These flies live around a third longer than normal and are especially resistant to heat, stress, and starvation.

In mice, removing a gene called *p66shc* seems to make the cells of the mice more resistant to free radicals, and also lengthens life by about 30% with no obvious side effects. Biologist Steve Austad at the University of Idaho is collaborating with a lab which has lengthened the lives of another strain of mice by genetically engineering them to produce more of the antioxidant catalase, specifically inside their mitochondria.

Most importantly, there are at least two drugs that have now been found to lengthen the lives of animals and which seem to have antioxidant properties. Biologists Simon Melov and Gordon Lithgow at the Buck Institute for Age Research found that two drugs that mimic the effects of antioxidants inside the body could lengthen the lives of nematodes. The drugs, EUK-8 and EUK-134, act like SOD and catalase inside of cells. They're manufactured by a private company called Eukaryon. When Lithgow and Melov fed the drugs to nematodes, the nematodes lived an average of 44% longer. They're currently testing these drugs on mice as well. Another drug called 4-phenyl butyrate that protects fruit flies against the effects of free radicals also lengthens their lives.

All told, there's a large and growing body of evidence that increasing the body's defenses against free radicals can extend life. The first evidence that lifespan could be increased this way came in 1999. In the few years since then, researchers have already made impressive progress. Their successes are sure to attract other researchers to the field. If the pace of research continues, we may see therapies to increase our lifespan enter human trials in the next decade.

Chapter 5 – Designer Lifespans

While geneticists have been looking at the use of genetic engineering to prolong youth, another group of researchers have found a simple and reliable way to lengthen life: Feed animals less food while maintaining their full intake of vitamins and other essential nutrients. This is what scientists call caloric restriction, or CR. For reasons we've just recently begun to understand, CR lengthens the lives of mice, rats, nematodes, fruit flies, spiders, guppies, dogs, chickens, and plenty of other species. There's preliminary evidence that it works in humans as well. When started at childhood in animals it usually lengthens life by 30 percent to 40 percent. For a typical American, that would add up to thirty years to life. That would mean an average life expectancy of 107 instead of 77, and a maximum lifespan around 170 instead of 120.

On the other hand, imagine *always* choosing the salad instead of the double bacon cheeseburger or never eating dessert for the rest of your life. Most of us already know that cutting calories back to *reasonable* levels is good for us. Despite that, obesity is booming in the developed world. Caloric restriction has far greater benefits than just eating a reasonable diet. No reasonable diet will extend your life 40 percent beyond the human norm. No reasonable diet will keep you healthy and fit decades longer than your free-feeding peers. But caloric restriction is also much more severe than a reasonable diet. It requires not just discipline, but intensive planning of your meals. It forces you to pay careful attention to your calorie and nutrient intake. Most of us just don't have the will or the time put ourselves on a CR diet.

Scientists researching CR know this full well. They have no illusions about the human willpower. They also know that humans will buy products that keep them young or that *keep them thin*. Inspired by this opportunity, some prominent and reputable caloric restriction researchers are now searching for caloric restriction *mimetics* – drugs that mimic the effects of CR on the body, while still letting you eat what you want.

Researchers have known for almost a century that reducing calories in lab animals could have health benefits. Francis Rous is best known for winning the Nobel prize in 1966 for his work on cancer. Far earlier, in 1914, he showed that reducing the food intake of rodents substantially reduced their incidence of tumors. Over the following decades, researchers slowly expanded this picture. In 1930, Clive McCay and his colleagues at Cornell University showed that rats who had their food restricted lived longer than rats that ate all they wanted. By the 1960s, the case for caloric restriction in animals was solid. Scientists all over the world had reproduced CR's life-extending effects on several different animals.

Scientists have now studied CR on a wide variety of other creatures, including rats, fish, cows, and dogs. In almost all of these species, caloric restriction started in childhood extends life by 30 to 40 percent. Starting it by early middle age extends life by about 20 percent. The numbers vary from species to species, but most fall inside this range.

Exactly how and why caloric restriction works is not yet clear. Increasingly, however, researchers believe that it taps into the same mechanisms that genetic changes that slow aging affect. In particular, organisms that consume less food seem to experience a slower rate of damage to the DNA inside their cells. This may be a result of producing fewer free radicals – a cell that is provided less energy will produce fewer of the waste products that damage it. Or it may be for other reasons not yet discovered. Either way, caloric restriction does more than make the lives of animals longer – it seems to keep them young and healthy longer.

Calorically restricted mice are more physically active than their free-feeding peers, and maintain that activity later into life. Biologist Steve Austad says, "One of the striking things that you don't often read about these calorically restricted animals is they have enormously increased spontaneous levels of activity; whereas, a young mouse might run a kilometer in a night, a caloric restricted mouse might run six or seven kilometers in a night. Whereas, a normal mouse would stop running at all by the age of eight months, these animals are still running several kilometers a night at the age of two years."

Caloric restriction also protects the mind against the ravages of age. After adulthood, lab animals show a slow decline in memory and other mental functions, just like humans. Lab animals on CR diets still show this decline, but it's remarkably slowed. Studies have shown that old mice on CR diets do better on memory and other mental tests than mice of the same age who've been allowed to eat whatever they want. In 2002, Christiaan Leeuwenburgh at the University of Florida showed that caloric restriction slows the rate of death of neurons in the brain, possibly explaining why CR animals stay mentally fit.

CR staves off disease as well. Calorically restricted animals have a lower risk of cancer, heart disease, diabetes, and just about every other chronic, life-threatening disease. CR animals are also more resistant to stress. They can survive toxin levels, heat levels, and surgeries that other animals can't. Their immune systems decline more slowly than do the immune systems of free-feeding animals.

Calorically restricted animals *look* younger too, outside and in. They stay sleek and trim. Their hair stays where it's supposed to and retains the color it had in youth. Inside they have the biochemistry of younger animals. As animals age, their levels of glucose and insulin in the blood rise, and their levels of a hormone called DHEAS decline. Old calorically restricted animals have lower blood glucose and insulin than normally fed animals, and higher DHEAS.

One of the most interesting effects of caloric restriction is what researchers call compressed morbidity – a shortening of the period of ill health at the end of life. CR animals often seem to live their lives in good health, free from major diseases, and then just keel over one day, long after their free-feeding peers have all died. This could have profound implications for human society if CR-based techniques work for us as well. Rather than a long period of infirmity at the end of life, with its attendant medical costs and burdens on loved ones, life extended humans might more frequently die in the prime of health, while still living as independent and active members of society.

The results above have been shown in a wide variety of animals. Just recently, scientists have started to find similar effects in primates, our closest relatives. Since 1987, the National Institute of Aging in Bethesda, MD has been studying the effect of caloric restriction on a colony of rhesus monkeys under the supervision of researchers George Roth and Mark Lane. In 1989 the University of Wisconsin at Madison started a similar study. All told there are about 300 primates being studied today between the two institutions. In addition to rhesus monkeys, the NIA is also studying caloric restriction on shorter-lived squirrel monkeys.

Rhesus monkeys live an average of about 24 years in the lab, with a maximum lifespan of 40 years. That's a shorter lifespan than humans, but much closer than the three years that mice live. Because rhesus monkeys live so long, though, it's going to take a long time to see the full effects that caloric restriction has on these animals.

In the less than two decades that have passed since the study began, though, the researchers *have* seen marked effects. In all of these cases the CR monkeys are compared to a control group who can eat whatever they want. Roth and Lane have found that the CR monkeys have lower insulin and blood glucose, just as we'd expect. They have higher levels of DHEAS – the hormone whose level normally declines with age. They have lower blood pressure and lower cholesterol. They have more flexible arteries, like younger animals. About half as many show any signs of heart disease, cancer, and diabetes than among the free-feeding monkeys. Most importantly, the death rate so far in the caloric restriction group has been just over half of the death rate in the free-feeding group. While the total number of deaths in the colony is too small to rule out chance, the lower rate of death is also consistent with all the other findings. In short, caloric restriction seems to be keeping these primates young.

Roth and Lane have also shown that some of the biochemical changes that come with caloric restriction in animals are associated with longer life in humans. Calorically restricted monkeys have lower body temperature, lower insulin levels, and higher levels of DHEAS for their age than the norm. Each of these traits is associated with longevity in humans in the Baltimore Longitudinal Study of Aging, the most comprehensive study of aging ever conducted in humans. In addition, humans who practice caloric restriction show similar biochemical changes, suggesting that CR affects humans in the same way that it affects other animals. Put it all together and there's a good case to be made that caloric restriction is likely to work in humans.

There are drawbacks to CR as well. Calorically restricted animals have less fat to insulate them and lower core body temperatures, making them more vulnerable to cold. CR animals are less fertile and less interested in mating than non-CR animals. They're smaller and less muscular than animals that eat all they want. And they may have weaker immune systems when they're young (though their immune systems stay healthy longer).

Humans who have voluntarily tried caloric restriction tell similar tales. Almost all of them report a loss of libido. Most are cold at temperatures that others find comfortable. Some report bouts of shivering. And many say that they fantasize about food constantly. Is a longer life worth it, if it comes with side effects like these?

If CR does work in humans and we can somehow eliminate the side effects, it may still remain unpopular. Most of us just aren't going to subject ourselves to the lifelong denial of the plentiful food that surrounds us. So to bring its benefits to humans, researchers will need to develop drugs that mimic the effects of CR while allowing us to eat all we want. These drugs will also need to avoid the side effects of CR. Achieving that combination in a single drug will be challenging, but there's reason to believe it's doable.

In the film *Monty Python and the Holy Grail*, there's a scene where an old man is thrown onto a cart of dead plague victims. "I'm not Dead Yet!" cries the still living geezer, but his being alive is an inconvenience to those who just want him to hurry up and die. Biologists Blanka Rogina and Stephen Helfand at the University of Connecticut had the same problem with fruit flies in 2000. That year Rogina and Helfand were running an experiment on genetically modified fruit flies. The experiment had nothing to do with aging or life extension. But as time passed, some of the flies just weren't dying when they were supposed to. The researchers were impatient - all the flies had to die before they could finish the experiment. Then the realization hit - they'd found a gene that was extending the life of these flies.

Helfand and Rogina named the gene *INDY*, short for I'm Not Dead Yet, in honor of the old man from Monty Python. Flies that carry a specific alteration of the *INDY* gene have a flaw in their metabolism. They can't quite get all the energy out of the food they consume. This puts them in a state similar to caloric restriction. *INDY* flies live almost twice as long as other flies on average. As far as the researchers can tell, the *INDY* flies were just as active, just as large, and just as healthy as normal flies.

So far the researchers have found only one disadvantage to the *INDY* flies. When they're fed all they want, the flies are more fertile than their unmodified cousins. But when food is scarce, they're noticeably *less* fertile. This makes sense - because their metabolism is inefficient, they need to eat more to stay healthy. Conversely, they're more sensitive to starvation than flies that get more energy out of each morsel they eat.

The finding of the *INDY* fly points to a possible mechanism for a life extension drug in humans. Humans carry a gene very similar to *INDY* that's involved in metabolism. A drug that targets this gene might cause the same effects as those seen in the *INDY* flies.

Rogina has also collaborated with geneticist Stewart Frankel at Yale. In a study related to *INDY*, Frankel and Rogina found that flies with a genetic mutation that reduced the level of an enzyme called Rpd3 lived between 33 percent and 50 percent longer than normal flies. Like INDY, Rpd3 is involved in metabolism. More importantly, a drug that researchers think affects Rpd3 has already been shown to lengthen the lives of fruit flies. In 2002, Hyung-Lyun Kang at the National Institutes of Health showed that feeding fruit flies a drug called 4-phenyl butyrate extended their lives by about a 40 percent. These examples raise the possibility that in humans a drug, not just gene therapy or caloric restriction, could be used to slow aging.

At least three other teams are currently working on drugs that mimic the effects of caloric restriction. Roth, Lane, and neurobiologist Donald Ingram published an article in 2002 in *Scientific American* describing their search for a drug that extends life by partially blocking the use of sugar inside the cell. They're currently working with a drug called 2-deoxy-D-glucose (2DG). 2DG is too dangerous for human use – it becomes toxic if too much is taken - but at low, carefully measured doses it may have life extending effects. The researchers are currently feeding 2DG to laboratory rats. The rats are showing some of the signs of caloric restriction – lower blood glucose, lower insulin, lower body temperature, and slightly reduced weight – even though they eat all they want. In another year or so we'll know whether these rats actually lived longer than the controls. If the rats *do* live longer, then we'll have evidence that a drugs can in fact mimic the effects of caloric restriction. Roth has formed a company called GeroTech that hopes to bring these products to the market. And he's not the only one. "There are another dozen companies working on CR mimetics, most of them using different mechanisms" he says.

One of those companies is Boston based Biomarker Pharma. Biomarker was founded by biologist Stephen Spindler. Biomarker's approach is to find the genes that become more active in animals that are placed on a CR diet, and use these as targets for drugs.

Spindler and his colleagues at Biomarker are studying the genes that are expressed differently in animals on caloric restriction diets – the genes that become more or less active when an animal goes on CR. They've already found that most of these gene expression changes happen very quickly – within just a few weeks of starting a CR diet. Using this information, they're looking for drugs that mimic the effects of these genes on the body. Basically, they have a systematic way to find new drugs like 4-phenyl butyrate and 2DG.

So we've seen that caloric restriction works in almost every animal its been tried in. It seems to be working in monkeys, our close relatives. And various studies in humans suggest that it will probably, though not certainly, extend human life as well. Animals with genetic mutations that reduce the amount of energy they can get out of food seem to live longer, but without obvious signs of the side-effects that plague CR animals. Researchers have already found one drug that works on one of these genes and extends life. A second drug is producing physical changes similar to CR, and we'll soon know if it extends the life of rats. Using gene expression studies, scientists are likely to produce many more of these drugs in the coming years. We can't be sure if we'll ever find a drug that mimics CR and is safe for humans to use, but today the evidence suggests that it's possible.

The life extension techniques we've talked about all turn out to be related to each other in some way. Calorically restricted animals and animals with mutations to their IGF-1 receptors both seem to suffer less damage from free radicals. Caloric restriction, IGF-1 related genetic changes, and antioxidant related genetic changes all render animals more resistant to toxins, ultraviolet light, heat, and other stresses that can increase the number of free radicals in the body. And animals on caloric restriction diets have less IGF-1 in their bloodstreams.

It's also important to reiterate that all three of these techniques seem to preserve healthy, youthful life. Animals who have their lives extended by any of these techniques stay physically active to later ages than their peers who go untreated. They retain their ability to learn and remember to older ages than normal animals. They develop fewer tumors and less heart disease. In old age they look and act like younger animals. When they finally die, it's often without any signs of the cancers or stiff arteries or atrophied muscles that plague other old animals. As far as we can tell, all three of these techniques actually stretch out the prime of these animals lives.

Despite these common threads, the three techniques have important differences and can complement one another. In fact, studies that have combined caloric restriction with mutations in the IGF-1 receptor gene have already increased lifespan more than either method alone. Andrzej Bartke at Southern Illinois University showed that Ames dwarf mice, a breed of long-lived mice that lack growth hormone, live even longer when placed on caloric restriction. Ames dwarf mice live about 50 percent longer than normal when they can eat all they want. When placed on a caloric restriction diet, they live around 75 percent longer than normal. Researchers have found similar effects when they've put fruit flies with a mutation in their insulin-like receptor gene on caloric restriction. 75 percent life extension in a mammal is fairly incredible. If we could increase human life expectancy by the same amount, we'd see average human life span in the developed world reach 135 years, and maximum human lifespan go past 200 years.

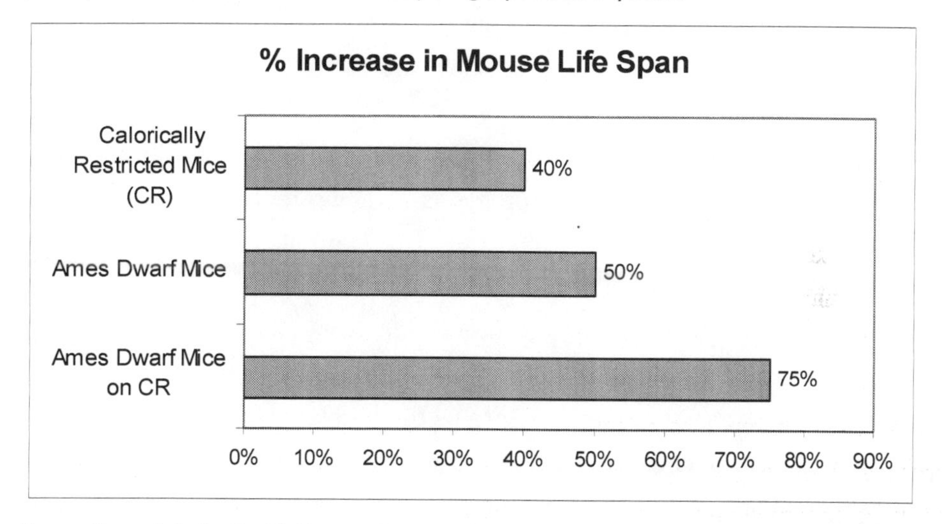

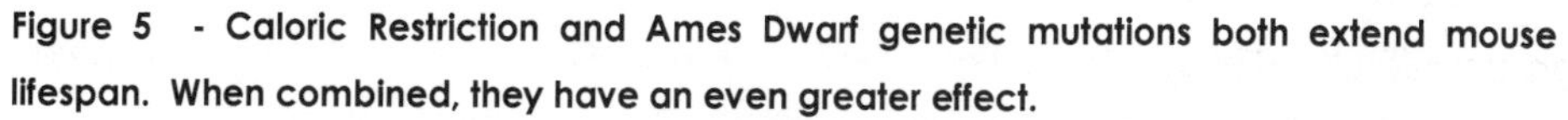

Figure 5 - Caloric Restriction and Ames Dwarf genetic mutations both extend mouse lifespan. When combined, they have an even greater effect.

So the insulin-like receptor techniques and caloric restriction work well together. What about the anti-oxidant techniques that guard against free radical damage? So far no one has published any results from experiments that combine free radical resistance genes with caloric restriction or insulin-like receptor mutations. Increasing lifespan by boosting resistance to free radicals is still too new a field. Today at least one research group is planning such an experiment though. In the coming years we'll find out whether all of these techniques can be combined to slow aging even further. If they can, the results may be even more striking.

The approaches to slowing the aging process we've talked about so far are based on experimental successes. They all take the basic approach of finding a technique that slows down aging in lab animals, and then making drugs that mimic that technique. Cambridge gerontologist Aubrey de Grey, one of the more controversial figures in the field of aging, believes that there are more effective techniques that could dramatically lengthen human life He says that "indefinite postponement of aging may be within sight."

This is an ambitious statement, yet de Grey is a respected thinker in the field. He chaired the 2003 meeting of the Institute of Applied Biogerontology – one of the more prestigious of the international aging conferences. He's also one of the signatories of the position paper that dismisses all current 'anti-aging' products on the market as scams. De Grey's optimism may stem from his unique viewpoint on the field. Originally trained as a computer scientist, he approaches the basic problems of aging more like an engineer than a biologist.

In a paper co-authored with respected aging researchers Bruce Ames, Andrzej Bartke, Judith Campisi, and others, de Grey laid out a multi-prong strategy that he believed would all but eliminate aging in humans. What's more, the same techniques could restore us to youthful vigor.

The most widely acknowledged of the problems de Grey focuses on is one we've already discussed – the slow decay of the mitochondrial powerhouses in our cells thanks to the constant barrage of free radicals. Mitochondria are particularly susceptible to damage because they carry their own DNA, but they lack the powerful defenses and repair abilities that they need to protect it.

99.95 percent of the DNA in each cell in your body is stored in the cell nucleus. The nucleus of a cell is a safe haven for DNA. Free radicals and other dangerous molecules are mostly kept out of it by a protective membrane. And inside the nucleus repair enzymes are constantly checking your DNA for problems and correcting them. Your mitochondria, on the other hand, reside outside the nucleus along with the rest of the cell's machinery. Most of the proteins needed to build mitochondria are located in the cell nucleus. Evolution has moved them there over time. But the job isn't quite done. The last 13 genes mitochondria need (out of about 1000) are still in the mitochondria themselves. And lacking the defenses and repair mechanisms of the nucleus, they're quite vulnerable to damage from free radicals and other sources.

De Grey proposes that we finish the job that evolution has started - moving the mitochondrial DNA into the cell nucleus, where it will be better protected against aging. The way to do this is by inserting the last 13 genes involved in creating mitochondria into our nuclear DNA. As de Grey points out, researchers have already moved one of those genes into the safe harbor of a cell nucleus while preserving all of the cells functions. As early as 1986, Australian biochemist Philip Nagley at Monash university succeeded in moving a mitochondrial gene in yeast into the nucleus with no apparent ill effects. While a great deal of work remains to move all thirteen genes, the principle has been demonstrated.

De Grey and his co-authors also propose a host of other techniques, like destroying garbage that builds up inside and outside of cells, and genetically engineering humans to eliminate the key genes that cancer depends on. With a focused researched effort, they believe it would be possible to use these techniques to rejuvenate aged mice and virtually eliminate their aging within 10 years. A result that impressive would have the world clamoring to adapt it to humans. But de Grey's targets are far beyond those of most other researchers in the field of aging. Should he be taken seriously?

De Grey is a controversial figure in aging research, but he's written some insightful and important papers. Some of his co-authors on the paper that presents his anti-aging strategy have impressive credentials themselves. Bruce Ames is one of the most cited biochemists in the world. In the last few years he's made his own controversial claims about the ability of some supplements to retard aging. Even so, there's no denying that he's made important contributions to biochemistry over the last forty years. Andrzej Bartke discovered one of the genes that extends lifespan in mice, and also showed that combining that gene with caloric restriction could extend life still further. Judith Campisi heads the department of cellular and molecular biology at the Berkeley National Laboratory and has done important work on the links between aging and cancer.

These are serious scientists who've contributed to the field of aging. Have they gone off the deep end in this proposal? De Grey's proposals make sense on paper. He makes a plausible argument for a set of deep changes we could make to the human body to eliminate the major causes of aging that we know of. The major case against de Grey's proposal is that it's entirely theoretical at this point. No one has shown that the techniques he proposes lengthen life in animals. In general, biology is an extremely messy, complicated field. What looks like right on paper doesn't always work in the lab or in the body.

The best way to find out if de Grey's proposals make sense is to try them. He's right about that. He's also right that each of the techniques he proposes should be possible in lab animals right now. De Grey is currently attempting to create a research institute to fund and coordinate research along these lines. If he's successful, then we'll know in about ten years whether or not his ideas actually work.

Whether or not de Grey succeeds in his more ambitious goal of halting ging, there's good reason to hope that we'll see safe and effective age-slowing therapies in the next few decades. Caloric restriction seems likely to work in humans and researchers are closing in on both genes and drugs that can mimic CR's effects. The sudden flurry of discoveries of new genes that slow aging in lab animals is also just the very beginning. In the next few years we'll see even more genetic discoveries, and further attempts to combine multiple techniques to lengthen life even further. It's not too hard to imagine a strain of mice that live twice as long as normal. And with every genetic discovery there will be additional avenues for biotech and pharmaceutical companies to pursue in creating drugs that slow the aging process.

The leading researchers in extending the lives of animals seem to agree that we'll soon be able to do the same to humans. Prominent and well-respected researchers like Cynthia Kenyon, Tom Perls, Lenny Guarentee, George Roth, and Mark Lane have already expressed their beliefs in a very tangible way: by founding companies to bring their discoveries of life extension in animals to human beings. As Kenyon says about extending human life "The lights are green everywhere you go."

Michael Rose, an evolutionary biologist who created super-flies that live twice as long as normal with greater vigor and endurance, thinks scientists have all but proven that they can manipulate human aging. "Aging is not immutable. It's not God's grace. It is a genetic problem, and you can solve it."

Steve Austad, whose work has been focused on studying the evolution of aging rather than finding specific ways to lengthen life, is also optimistic about bringing the science to humans. "The sorts of therapies that I've been talking about that work in animals [will] actually get extended to humans, and I think that some of them will end up being relevant," he says. "I think it's easily possible that we'll get a few additional decades of human life expectancy." At the same time, he's a bit more conservative on when humans will benefit. In 2002 he said, "I think the time horizon for anti-aging therapies [in humans] is probably something on the order of 30 to 60 years from now."

In contrast, Tom Johnson, the pioneering researcher who first showed that small genetic changes could dramatically lengthen the lives of laboratory animals, has suggested that we'll see effective life-extension drugs on the market by 2020.

Judith Campisi is even more bullish, "We know we can extend the life span of mammals. There is no reason to believe that we couldn't do the same today in humans."

Combine all the evidence and the odds look good that some day soon we'll see a drug or gene therapy that slows the aging process in many different kinds of animals, and that seems safe enough for human trials. A person who started using such a therapy while in their twenties or thirties could have decades added to their life. Based on the evidence from animal studies, those years would vital ones – years where our bodies looked younger, where our minds held onto the flexibility of youth, and where we could live life to the fullest. Then the field of life extension will hit a whole new era – one in which we'll see the effects that longer lives have on our society.

Chapter 6 – Methuselah's World

If researchers succeed in developing drugs or gene therapies that can slow human aging, who will benefit? A few chapters ago we talked about the price of enhancement. Like other technologies, biotech enhancements will likely be expensive when first introduced, and then drop in price over time.

There's another, more specific reason to believe that life extension techniques will make it out to the masses - the historical record. Throughout most of human history, life expectancy hovered at a very low level - somewhere between eighteen and twenty-five years. Then around 1800, life expectancy began to rise quickly in developed countries - Europe, the United States, and Japan. For the next century, life expectancy rose in these countries while remaining low in the developing world. Eventually, though, sanitation, vaccination, and other public health techniques made it out to developing countries of Asia, South America, and Africa, starting a life extension revolution there. The gap between rich and poor countries has been largely a lag in time ever since.

In the 1800s the rich got a head start in increasing life expectancy, just as they may with new biotech enhancements. Over the last century, though, the life expectancy of still developing nations (as defined by the United Nations) has been growing faster than that of countries the UN defines as already developed. The length of the average life in still developing nations has climbed from twenty-eight years to sixty-three, and is on track to reach seventy-five years in 2050. Meanwhile life expectancy in developed countries has gone from fifty years in 1900 to seventy-five years in 2000, on its way to eighty-two in 2050.

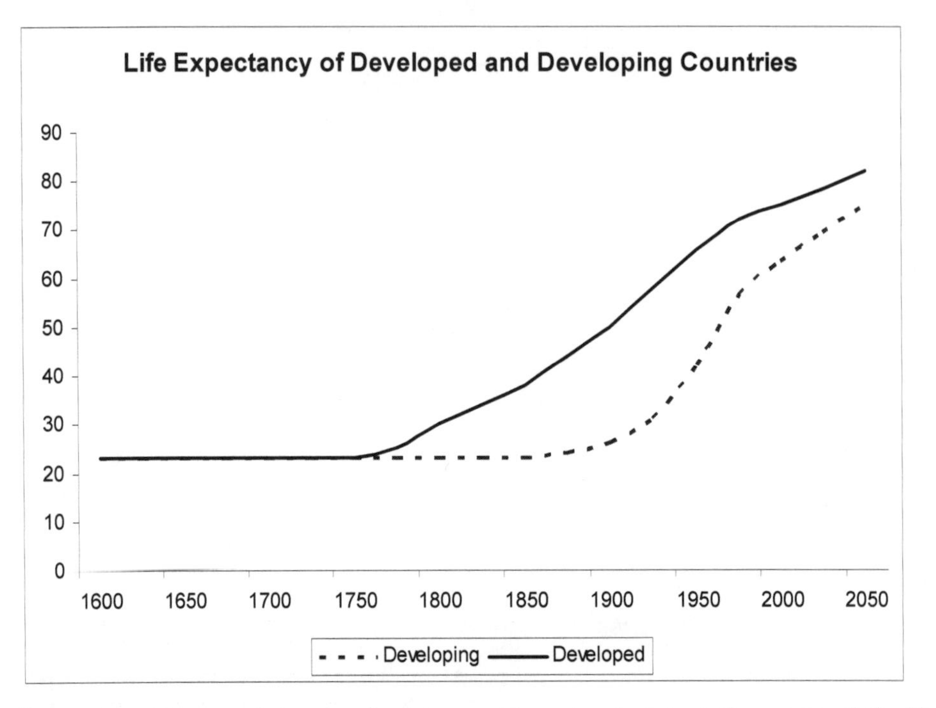

Figure 6 - Compared to developed countries, developing nations saw their life expectancies rise later but more quickly.

Those numbers show that the life expectancy gap between rich and poor is narrowing. In 1900, a child born in a developing country could expect to live, on average, just over half as long as one born in a developed country. By 2000 that ratio had climbed to 84 percent. By 2050 the UN Population Division expects life expectancies in less developed nations to reach 91 percent of those in developed countries. In the most concrete terms we can measure it the difference in material well being between rich and poor has shrunk over the last century, and seems set to shrink further over the coming fifty years.

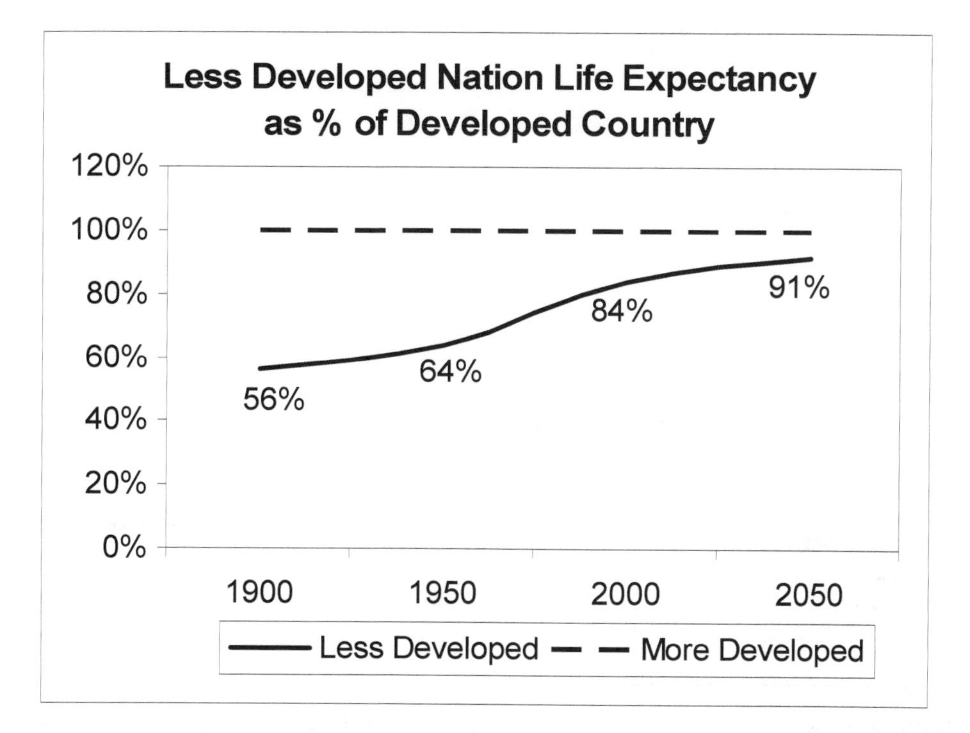

Figure 7 - Developing countries have closed much of the life expectancy gap with already developed nations over the last century.

Combined with the evidence for rapidly dropping price of information goods, the past history of life expectancy suggests that the poor of the world will eventually benefit from biotech enhancements, and from age-slowing techniques in particular. The poor won't have access to such techniques until some time after the rich, but once they do get access, they may make progress more quickly, closing the gap over time.

Dodging the Demographic Bullet

If age-slowing techniques do become widely available, they could help society avoid the problems of an aging population. In 1900, there were 3 million people over the age of sixty-five in the United States. By 1996 that number had grown to *35 million*, more than a factor of eleven. By 2030 it's expected to double again, to *70 million*. One in five Americans will be over the age of sixty-five. In other countries the shift will be even more dramatic.

In Japan, for instance, the median age of the population is expected to rise to fifty-three by the year 2050. In 1950, the median age was twenty-two. The average citizen of Japan will be almost two and a half times older in 2050 than at the end of World War II.

This demographic shift comes with consequences. In the United States, for example, as people hit retirement age, they'll leave the workforce, depleting countries' economic output and tax revenues. At the same time, they'll start to collect from Social Security and other pension funds. Overnight they'll go from being a net economic contributor to society to a net economic drain.

Social Security and other national pension plans are already facing a crisis brought on by this aging population. In 1950 in the U.S. there were more than 16 people of working age (twenty to sixty-four) for every person aged sixty-five or older. Today that ratio is close to three to one. By 2030 it's projected to reach just two working-age Americans per retirement-age American. If everyone retires at age sixty-five, there will simply not be enough workers generating money to pay for the pensions of the retirees.

What's more, the elderly consume radically more health care than the young. In developed countries, individuals over age sixty-five spend anywhere from three to five times what people under sixty-five spend on healthcare.

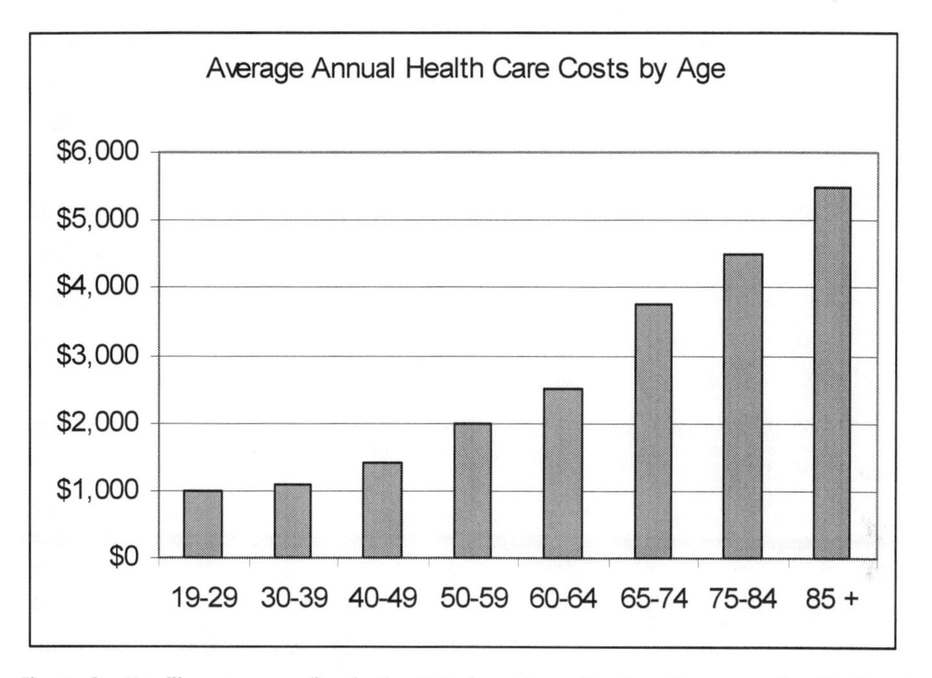

Figure 8 - Health care spending in the U.S. rises dramatically with age as the likelihood of serious ailments increases.

The total cost to society is staggering. In 2001, the over-sixty-five group accounted for almost half of all health care spending in the US, despite making up only about 13 percent of the population. Or more precisely, that much is *spent on them*. Most health care is paid for by insurance plans or government programs, both of which pass the costs on to younger, healthier individuals. As we invent newer, more expensive life-saving treatments, and as the population of developed countries continues to age, the costs will rise even further.

All told, the Congressional Budget Office projects that Social Security, Medicare, and Medicaid will rise from 7.5 percent of US GDP in 2000, to almost 15 percent of US GDP in 2030. That's one out of every seven dollars that every person in the US makes, going to fund programs for the elderly. And the lion's share of this growth isn't in Social Security – it's actually in the cost of Medicare for the old. By 2075 the CBO projects that these programs will require one of out every *five* dollars earned in the US, and their costs will still be rising.

This crisis will hit before life extension technologies have had much effect on the population. Governments will respond in a variety of ways, including raising the retirement age and reducing benefits. Ultimately, they may go further and link pension payments to ill health or other disability that prevents one from working. After all, when Social Security was first created, its stated purpose was to protect the old from the ravages of "penniless, helpless old age." A sixty-five year old who's fit enough to play tennis three times a week and sharp enough to re-enroll in college isn't what the founders of Social Security had in mind.

Slowing human aging is one of the few ways out of the Struldbruggian dilemma of an aging population. If the projections of Johnson, Kenyon, de Grey, and other scientists are correct, life extension technologies will start to make their impact around the time that big changes are due in national pensions. By keeping some of the elderly biologically younger, they'll slow the growth of medical spending on the aging population, combating the largest increase in spending the CBO projects.

In addition, life extension seems to compress the duration of frailty at the end of life. In the words of biologists, life extended animals display *compressed morbidity*. They frequently remain fit and active until their final days, then die quickly. Sometimes those deaths have no apparent cause. Autopsies of calorically restricted mice, for instance, find that about one third of the animals have nothing obviously wrong with them at the time of their death. With current medicine, on the other hand, the costs rise exponentially as the end of life nears. 28 percent of Medicare spending goes to patients in the last year of life. 47 percent of all Medicare spending happens in the last three years of life. Dying while still in good health would eliminate much of the cost of these final years, while also reducing suffering.

In addition to cost reduction, slowing aging would also inject dollars into the economy. Life extension techniques would increase the number of elderly who are still mentally and physically sharp enough to work and contribute the economy. Ironically, as a reward for good genes, good health habits, or good sense in pursuing age-slowing therapies, the fittest and healthiest older people will find themselves required to work for their money. Losing the promise of a long government-funded retirement isn't very pleasant. Even so, most of us would rather be in good health and have to work, than be in poor health and get social security checks.

Across the developed world, the current national pension systems are in trouble. Without some sort of reform, the Congressional Budget Office projects that Social Security, Medicare, and Medicaid combined will take up two thirds of the U.S. federal budget by 2030. By 2040, the Social Security trust fund will be empty. To maintain solvency, the U.S. and other nations will need to either raise taxes dramatically or reduce benefits. Given a healthier workforce, the most obvious step may be to delay the age of retirement. Slowing the rate of aging is one of the few steps that could help delay this demographic crunch.

Life Extension and Population

How will longer lives affect world population? Certainly anything that keeps people alive longer will increase the number alive at any given point. However, the details of population growth can be rather counterintuitive. Consider that today the countries with the longest life expectancies at birth have populations that are remaining steady or even *shrinking*. For example, the UN Population Division expects the populations of Japan, Italy, Germany, and Spain to all decrease over the next 50 years, despite the fact that Japan has the highest life expectancy of any large country, and the Western European nations are not far behind. At the other extreme, the countries with the fastest growing populations are those with relatively low life expectancies – countries like India, China, Pakistan, and Nigeria.

Throughout most of human history, birth rates and death rates were both extremely high. In the year 1000 AD, birth rates were around 70 births per 1000 people per year, and death rates were about 69.5 deaths per 1000 people per year. So each year a town of 1000 people would on average increase its population by half a person. That's a 0.05 percent rate of growth.

As societies advance, nutrition, sanitation, and medicine all serve to lower the death rate. When births outstrip deaths, population soars, as happened through the twentieth century. But in the last few decades the birth rate has also dropped sharply, particularly in the rich developed nations of the world. As countries gain in wealth and education, and especially as women in those countries gain greater rights, resources are increasingly devoted to education and career, rather than raising a family. This demographic trend is spreading from Europe, Japan, and North America to the rest of the world. Worldwide the birth rate is now 21 births per year per 1,000 people. The death rate is around 10 deaths per year per 1,000 people. So overall the world population is growing at roughly 1 percent, slower than at any time in the last few centuries.

Because the birth rate is double the death rate, phenomena that affect fertility can have a far larger impact than similar size effects on the death rate. For example, between 2000 and 2050, the UN expects around 3.7 billion people on earth to die, while another 6.6 billion are born. Cutting the death rate in half would increase population by around 1.9 billion people. Doubling the birth rate would increase population by a further 6.6 billion. Thus the birth rate is a more significant lever on the size of the population.

Projecting population growth into the future is a tricky thing. It depends on just how quickly economic growth improves healthcare, sanitation, and nutrition. It also depends on how quickly those important birth rates come down. The United Nations Population Division produces high, medium, and low estimates for future population. Birth rates and death rates differ by just a few percent between the scenarios, but over decades it adds up. In 2000, world population stood at 6 billion people. Looking forward, the UN predicts that in 2050, world population will be somewhere between 8 billion and 11 billion, with a "medium" prediction of 8.9 billion.

8.9 billion people is half again as many as are alive today. Yet in real terms it represents a growth rate about half as fast as the last 25 years. World population grew by 50 percent between 1975 and 2000. The next 50 percent increase is predicted to take until sometime after 2050, around twice as long. Ultimately the UN's "medium" model projects world population hitting a plateau of 10 billion people in 2100. Of course, projections 100 years in the future are perilous. They rely on a generally smooth progression of current trends. They don't take into account the possibility of global plagues, world wars, or technologies that radically alter mankind.

Life extension is just one of those technologies, yet its impact on population is surprisingly small. Demographer Jay Olshansky, despite being a pessimist about the prospects for slowing human aging, has shown that extending human life would have an incremental, rather than exponential, effect on population. "The bottom line is that if we achieved immortality today, the growth rate of the population would be less than what we observed during the post World War II baby boom", he says. If everyone were made completely immortal today, he calculates, and taking into account declining birth rates, global population would hit about 13 billion in 2100, rather than the 10 billion currently projected.

The immortality Olshansky is talking about isn't achievable. If we halted all aging, accidents, homicide, suicide, and infectious diseases would still kill people. Nor is complete halting of the aging process in our sights today. More likely, we'll achieve the technology to slow but not halt the rate of aging in the next 10 to 20 years. Then it will take additional years for people around the world to gain access to the technology. Even then, the impact on population won't be instant. Those who are already old won't benefit much from techniques that slow their aging rate. Instead, it will be the middle aged and young who reap most of the benefits. The combination of those factors suggests a gradual impact on worldwide population.

Here's a simple mathematical way to look at this. We can't predict exactly when life extension techniques will first be available or how quickly they'll spread, so we'll need to make some guesses. Imagine that life extension techniques first appear on the market in 2015. In the year after that (2016), let's say the overall death rate around the world drops by 1 percent as a result. And let's say that it decreases by another 1 percent each year thereafter, so that in 2017 age-slowing techniques have reduced the global death rate by 3 percent, and that by 2050 they've reduced it by 35 percent. That's quite an optimistic scenario. It implies that by 2050 we've managed to increase life expectancy in the developed world to around 120, and in the developing world to around 113 - an unprecedented rise in life expectancy, even compared to what's happened in the last two centuries.

If this extremely optimistic rise in life expectancy occurred, how large would the impact on global population be? If we simply reduce the number of deaths the UN forecasts each year by the percentages above (1 percent in 2016 climbing up to 35 percent in 2050), we get a total population in 2050 of 9.4 billion people, instead of the 8.9 billion the UN projects.

Five hundred million additional people are nothing to sneeze at, but as a proportion of the projected world population for 2050, it's less than 6 percent. To put this in context, that's less than the percentage change in world population between 1970 and 1973 - a noticeable change, to be sure, but not a catastrophic one.

It's also worth noting that some demographers believe that birth rates in the developing world will drop faster than the UN's middle scenario. Just a 5 percent lower birth rate yields a scenario where global population tops out at around 8 billion in 2050, and may come back down to below the year 2000 level of 6 billion in 2100. Achieving this is entirely plausible if the right resources are applied. A plateau and eventual reduction in global population depends on the spread of wealth, education, and freedom in the developing world. For example, the UN expects the population of Europe (not counting immigration) to shrink by about 0.4 percent per year between now and 2050. The overall population of the developed world (again not counting immigration) is expected to remain fairly constant. If the developing world can attain the present level of affluence of the developed world by 2050, we'll see overall world population level out and begin to drop.

Enhancement and Social Change

Health care, retirement, and population are all fairly crisp issues. They lend themselves to mathematical analysis. If people age this much more slowly, their health care costs will go down by so much. If people live this much longer, population will increase by so much.

Yet the prospect of conquering aging raises other, softer issues. In particular, how will longer lives affect social change and progress? Many human organizations are pyramidal. People struggle to get to the top, where they can wield real power. Once they get there, they don't want to let go. Business executives, incumbent politicians, and university department heads are just a few examples of this.

If human lives are lengthened, we can expect the people at the top to try to hold onto their positions. And in so doing they'll be robbing younger generations of opportunity. They may also serve to forestall innovation. In some academic fields, there's a saying that progress occurs one death at a time, as old stalwarts die, and a younger generation can finally get their ideas out. With older people clinging to life and power, this may not be possible. Indeed, we may see a sort of intergenerational conflict between the very old and traditional on one hand, and new generations willing to adopt more radical technologies and ideas on the other.

Social, political, and academic stagnation and intergenerational conflict are plausible risks. Yet there are also powerful forces working against stagnation of the old. Age-retarding technologies don't just extend life, after all. They keep our brains and bodies younger. Younger brains learn faster and adapt to new conditions more easily. A 65 year old in the middle of this century may have a brain as flexible as that of a 25 year old, yet with all the accumulated experience of 65 years of life.

A second consideration is our increasing ability to sculpt our own minds to boost learning rate or memory, or even to alter facets of personality. Competitive pressures will fuel the adoption of mind enhancing and sculpting technologies among the life extended. It's one thing for a manager or business leader to become stuck in his ways. It's another thing for a whole company or university to become stagnant. When that happens, other more flexible competitors will swoop in to take their slice of the pie. At an organizational level, there's a kind of natural selection. Organizations that are slow to change will die out. The nimble ones will probably be those with nimble people in them.

An older population is likely to trigger other changes in society. For example, in general the older you are, the more likely you are to vote. In the US election of 2000, only 34 percent of citizens aged 18-24 voted, while 73 percent of citizens aged 65 - 74 voted. In addition, older people are far less likely to commit violent crimes. The large majority of violent crimes are committed by young men aged 18 - 24. Polls have also found that the elderly are also less likely to support military action as a way to solve international conflicts than the general population. Combined, these suggest that an older population is both more engaged in civic life and more restrained in the use of violence. How that will impact society in the long term is difficult to say, but with the aging of the population we will likely find out, life extension or not.

In the end, the practical benefits of life extension seem to outweigh the risks. We know that slowing aging would be the most effective way to combat a wide variety of diseases, that it would soften the blow of the graying of society, and that it will have a surprisingly small impact on global population. Given those more concrete data points, it seems most reasonable to take a wait and see attitude towards the secondary effects that life extension will have on society. Undoubtedly longer lives will cause unforeseen problems, and as a society we'll have to come together to solve them, just as we've solved the problems that have come with previous social goods.

Ultimately, longer lives can't be considered in a vacuum. Life extension is just one of several transformative technologies humanity is developing. The other technologies discussed in this book will give men and women greater ability to remove their present limitations, to reinvent themselves, to learn and improve beyond the plateaus we now encounter. As we learn to redesign ourselves, we'll reach states of mental and emotional capacity for growth that simply can't be satisfied in one present human lifetime.

When we contemplate the three years that a mouse may live, we don't mourn its short time on this earth. In three years a mouse lives and learns as much as it's able, and more years wouldn't add meaning or quality to its life. Today a human lifespan may provide enough years for a man or woman to learn and grow as much as we're able. But in the decades to come we'll increase our capacity to learn, grow, and change over time. Eventually one hundred years may seem like a brief adolescence - an early phase of life, in which we learn many things but only barely reach maturity.

Today we mourn those who die in adolescence, because they have so much left to see and do in their lives. As we push back the limits on our minds and bodies, as we increase our own ability to grow and learn, we'll feel the same way about those who die at eighty or one hundred. With the power to transform ourselves our futures become less limited by our starting point, and more by the lengths of our lives. By extending those lives we'll give ourselves, both as individuals and as a society, room to grow and blossom in ways that we have yet to imagine.

Chapter 7 – A Child of One's Own

In the late 1960s, Lesley and John Brown, a young couple from Bristol England, decided to try to have a baby. 9 years later, Lesley had still not conceived. As it turns out, Lesley's Fallopian tubes were blocked. Her eggs couldn't pass down from her ovaries to meet her husband's sperm and fuse to form a single-celled embryo. Over those 9 years, John and Lesley saw doctor after doctor, none of whom were able to help.

In 1976 the Browns were referred to Patrick Steptoe. Steptoe, a gynecologist at Oldham General Hospital, was ten years into his own quest into new ways to help infertile couples through a technique called in vitro fertilization (IVF). With his colleague Robert Edwards at Cambridge, he'd been able to combine sperm and egg in a lab dish and produce early embryos. They'd implanted these embryos into their prospective mothers, but as yet none of the pregnancies had lasted longer than a few weeks.

Still, Steptoe and Edwards were optimistic. In their previous attempts they'd implanted the embryo into the prospective mother's body five days after conception, when it had reached a size of sixty-four cells. They theorized that by implanting the embryo into the mother's body earlier, when it was only eight cells, they might improve the odds of success.

On November 10 of 1977, Steptoe used a probe called a laparoscope to remove an egg from one of Lesley Brown's ovaries. Edwards mixed the egg with some of John Brown's sperm. One of those sperm penetrated the wall of Lesley's egg. Conception had occurred.

Two days later, Steptoe carefully inserted the growing embryo into Lesley's uterus. This was a tricky point in the process. In natural conception more than half of all embryos at this stage are flushed from the body without ever implanting in the uterine wall. But luck won out, and the embryo implanted in Lesley's body. She was pregnant.

Eight and a half months later, on July 25, 1978, Lesley Brown gave birth to a healthy, blue-eyed, blond-haired baby girl. Louise Joy Brown, the world's first "test tube baby" was born. Through IVF, science had given us the ability to create – or at least conceive - life in the laboratory.

The initial public response to IVF wasn't pretty. Jeremy Rifkin and other critics of biotechnology (including Leon Kass, now chairman of the President's Council on Bio Ethics) criticized the technique. Edwards and Steptoe were accused of playing God. After their second successful IVF delivery, public protest forced Steptoe and Edwards to halt their work for two years.

After the initial shock lessened, however, the technique rapidly gained in popularity. Since 1978, more than a million babies conceived by IVF have been born. More than 100,000 more are born each year. One out of every one hundred births in the U.S. is conceived through IVF. In the United Kingdom, France, and other parts of Europe where the cost of the procedure is more frequently subsidized, the number is one in every fifty births. In Australia, one in every *twenty* births is conceived through IVF.

The process is still essentially the same as when John and Lesley Brown went through it. Doctors take several eggs from the prospective mother and sperm from the prospective father and mix them in a lab dish. If all goes well, a handful of the eggs will be fertilized by sperm and become embryos. Between forty-eight and seventy-two hours later, two or three of the healthiest looking embryos, each of which has between four and eight cells, are implanted back into the mother to begin her pregnancy.

IVF may sound simple, but it's actually a rather challenging procedure. First, there is the cost. A single IVF cycle can cost anywhere from $5,000 to $15,000. Then there is the trauma to the mother. A woman going through an IVF cycle faces a daunting array of injections, tests, and procedures to determine her health, promote production of several eggs, extract those eggs, and implant the fertilized embryo(s) back into her uterus. All in all it's a physically, emotionally, and financially grueling process stretched out over at least two weeks. And there are no guarantees that a mother will get pregnant in a single cycle. Often the embryos fail to implant in her uterus, and the entire process must be repeated.

Despite this, IVF has become mainstream. The reason is very simple. Worldwide, as many as one in every six couples is infertile. Without IVF and other assisted reproduction techniques, they cannot have biological children, and the urge to have children is one of the most powerful instincts that we carry. It's embedded deep in our genes, and in the genes of every other species that's ever lived on this planet.

Faced with the choice of having no biological children or going through an expensive, painful, and uncertain procedure in order to have children, more and more couples are choosing the latter. This basic desire to reproduce fuels IVF, and is now fueling other techniques aimed at bringing healthy babies into this world.

Susan and Chris Dunthorne had a different problem than the Louise Joy Brown's parents. They weren't infertile – the birth of their son Joshua proved that. But Joshua died just four months later, a victim of cystic fibrosis.

Cystic fibrosis is a genetic disease. It's caused by a mutation in a gene called CFTR on the seventh chromosome. It's a recessive disease, meaning that it only manifests in children who inherit two broken copies of the CFTR gene. About one in every twenty Americans carries one broken copy of this gene. Thirty thousand people in the U.S. carry two broken copies and suffer from the disease.

The Dunthornes were heartbroken at the death of their son. They desperately wanted another child, but couldn't conscience the risk of passing on cystic fibrosis again. Either watching another child die, or watching him or her grow up afflicted with the disease promised to be too painful. As a result, they feared they'd never be able to have a child of their own.

That was before the Dunthornes learned about pre-implantation genetic diagnosis, or PGD. PGD is a step added to in-vitro fertilization. It was first developed to increase the success rate of IVF, by helping doctors pick the healthiest embryos – the ones most likely to implant in their mother's uterus and lead to a successful pregnancy.

In PGD, mother and doctor go through the normal steps of IVF. After embryos are conceived outside the body, they're allowed to develop for two or three days. At this point, when the embryos have reached the eight cell stage, doctors remove one or two of these cells and test them for the presence of healthy chromosomes. The other six or seven cells go on just fine without the ones that have been removed. At this early stage, all the cells in the embryo are identical, and each one is capable of producing any of the hundreds of kinds of cells found in the finished body.

The tests run as part of PGD can spot the difference between healthy embryos that are likely to lead to pregnancy, and chromosomally malformed embryos that are more likely to be flushed out of the body or lead to a miscarriage. By spotting major problems in chromosome structure, PGD has already proven successful at increasing the success rate of IVF, especially among couples with a history of many IVF failures.

PGD can *also* been used to spot genetic diseases. Once doctors have genetic material from the embryo, they can test it for the presence of any gene that has been identified. In the last decade parents and doctors have used PGD to test for Down's syndrome, sickle-cell anemia, muscular dystrophy, Huntington's disease, and other genetic disorders. All in all, several thousand PGD procedures have now been performed since the technique was first reported in 1989.

For the Dunthornes, PGD was a new hope. Their physicians at Hammersmith Hospital in London withdrew fifteen eggs from Susan's ovaries. After being combined with Chris's sperm, six of the eggs developed into embryos. Their doctors removed one cell from each of the embryos for testing. The first round of tests failed, and hope almost collapsed. Then the second round succeeded, showing that of the six embryos, two were healthy and free from the CFTR mutations that would lead to cystic fibrosis. Both embryos were implanted in Susan's uterus and one of them attached itself with her. Ten days later they got the news - Susan was pregnant. Nine months later their baby son Ethan was born, free of cystic fibrosis.

This is just the beginning for PGD. The new frontier is in using the technique to spot genes that don't guarantee a genetic disorder, but that increase the risk of some later disease. For example, PGD has been used to spot dangerous mutations of the BRCA1 gene, which increase the likelihood of cancer. It's been proposed to spot dangerous variants of the ApoE gene that increase the likelihood of Alzheimer's disease. Several PGD centers in the U.S. will now attempt PGD for any condition for which there is a genetic test – a list that is growing day by day.

PGD does have major limitations, though. You have thirty thousand genes. You differ from other humans in as many as 3 million ways. Yet IVF procedures typically produce only a handful of embryos. None of those embryos will ever be perfect. Indeed, the more we know about genetics, the more obvious the imperfections will be.

In Chris and Susan's case a different outcome of the genetic lottery might not have worked out so well. Of the six embryos they conceived, only two were healthy and free of cystic fibrosis. What if all of the embryos they'd conceived had carried the CFTR genes? Or a gene with a high risk factor for Alzheimer's disease? Or Parkinson's? If every embryo a couple conceives carries a high risk factor for *some* disease, what can they do?

There are in fact couples for whom this will always be the case. For example, there are three variants, or alleles, of the APOE gene - e2, e3, and e4. People who carry one copy of the e4 allele have a greater risk of both Alzheimer's disease and high cholesterol. People who carry two copies of the e4 allele have a higher risk - a 60% chance of contracting Alzheimer's in their lifetime. Every child inherits one copy of each gene from their mother and the other copy from their father. If either a man or a woman carries two copies of the e4 variant of APOE, then all of their children will be guaranteed to inherit it at least one copy. If *both* mother and father carry two copies of e4, then all of their children will also have two copies.

For parents like these, no amount of genetic screening can remove the risk factors from their offspring. Today such parents have two options. They can have a child and hope that medicine solves the problem by the time he or she is grown up. Or they can elect to not have a biological child, either adopting or simply not raising children at all.

Eventually, though, some couple won't be satisfied with these options. They'll ask their doctor "Isn't there something more we can do? Can't we just tweak this one little gene? Can't we change that dangerous e4 into a healthy e2?" Thousands of prospective parents will ask questions like this, to which their physicians will reply "no". Until one day some sufficiently motivated parents find a doctor who answers "Yes. We can tweak that one little gene."

To date, no one has altered the genes of a human embryo. The closest researchers have come to this is a rhesus monkey named ANDi. ANDi was the first genetically engineered primate brought into this world. Her creators took a rhesus monkey egg and inserted a jellyfish gene – one that makes jellyfish grow green. The gene is in every cell in ANDi's body. By inserting a foreign gene into an embryo of such a close relative of ours, scientists have demonstrated that genetically engineering human embryos is nearly feasible.

The technique for genetically engineering an embryo is much the same as for the kind of gene therapy we discussed earlier in the book. Scientists use a gene vector to deliver a new gene to the embryo in essentially the same way they would deliver a new gene to any other cell. Scientifically and technologically, the two techniques are almost the same.

There are important differences in the results produced when a gene is inserted at such an early stage, though. To illustrate these differences, let's compare ANDi, the genetically engineered monkey we just talked about, and Ashanti De Silva, the first human to receive gene therapy.

ANDi went through *germline genetic engineering.* She had a new gene inserted into her when she was just a single cell. That gene integrated itself into the DNA of that single cell. From then on, every time that cell divided to make new copies of itself, the new DNA was copied too. Every cell in ANDi's body carries the jellyfish protein gene. If ANDi has children, she has a 50% chance of passing that gene onto them as well. Her germline – the genetic information she can convey to future generations - has been engineered.

Ashanti, on the other hand, went through what's called *somatic gene therapy*. She was four years old when it happened. By this time Ashanti's body was made up of trillions of cells. Delivering the new gene to every one of her cells was effectively impossible. So researchers focused on just the most critical cells for her disease –blood cells. And because the process of getting genes into cells isn't perfect, only *some* of these blood cells carry the corrected ADA gene that Ashanti needs. Importantly, only her somatic cells (the ones that aren't involved in reproduction) were affected. If Ashanti has children, they'll inherit their genes from among the pool that Ashanti was born with, not the version that French Anderson and his colleagues injected her with.

Interestingly, altering the genes of a person before he or she is born is in some ways easier than altering them after birth. It's easier to get a gene into every one of a small number of cells than a large number. And at early stages of development, there's no immune system in the embryo to respond in potentially dangerous ways to the gene vector. So as medical motivations drive improvements to gene therapy, they will also improve germline genetic engineering. As researchers increase the safety and effectiveness of somatic gene therapy through new and better gene vectors, they'll be indirectly making genetic engineering safer and more effective as well.

There are already proposals for procedures that would blur the line between somatic gene therapy and germline engineering. In 1998, French Anderson proposed *in-utero* gene therapy. His idea was to perform gene therapy on a few-week to few-month old fetus suffering from a genetic ailment, while the fetus was still in its mother's womb. The Recombinant Advisory Commission, the watchdog that looks over all use of genetic technology in the United States, turned down the request. They were concerned that if gene therapy were performed during pre-natal development, the inserted genes might find their way to the child's sex organs, and then be passed down to future generations.

Despite this, there are medical reasons to think about in-utero gene therapy. Most children are still conceived the old fashioned way, through sex rather than IVF. In these children there's no chance to do genetic screening or alteration before implanting them in the womb. But around 3% of all pregnancies result in a child with a birth defect of one sort or other. The most common of these are congenital heart defects, to which several genes have been found to contribute.

Increasingly, concern around the health of unborn children has led to prenatal genetic testing. In the US around 13% of all expecting mothers go through some form of prenatal genetic test during pregnancy. Amniocentesis, the best known of these procedures, is now recommended for any expecting mother over the age of 35, or for those with a history of Down Syndrome or other genetic disorder in their family.

In January of 2004, a team of researchers and physicians from the University of California, San Francisco and Washington University published a paper in the British journal The Lancet, recommending that all mothers undergo amniocentesis or another prenatal genetic tests. The team, led by Miriam Kuppermann at UCSF, grounded their argument in economic terms. Amniocentesis has a cost and it also carries with it a risk of miscarriage between one quarter and one half of a percent. On the other hand it can help identify Down Syndrome and other problems early in a pregnancy. The total savings to society by early detection of fetal illness, Kuppermann showed, outweighed the cost and risk of the procedure.

What's more, newer tests like CVS (chorionic villus sampling) are safer, less expensive, and can be performed earlier in the pregnancy. These tests can pinpoint quite a few genetic diseases, including some that do irreversible harm before birth. For example, Tay-Sachs disease does permanent damage to the brain during gestation. In the near future prenatal tests will be able to detect it while a fetus is just a few weeks old.

By performing gene therapy early in the pregnancy, doctors could potentially eliminate Tay-Sachs, cystic fibrosis, congenital heart problems, or other genetic diseases before they have a chance to harm the child. And because the embryo is still rapidly developing, any genes inserted during the pregnancy will be spread through more of the body. In effect, the earlier gene therapy is performed, the easier it is.

For now the Recombinant Advisory Commission's ruling effectively forbids in-utero gene therapy. Yet as genetic testing becomes faster, easier, and more common, more prospective parents will find out that their unborn children bear some genetic disease. Those parents will be looking for ways to cure those disease.

In utero gene therapy offers such a route. And despite widespread unease about genetic technologies, most Americans seem willing to employ them for medical uses. In a 2002 survey of more than twelve hundred people, the Johns Hopkins University Genetics and Public Policy Center found that 73% approved of the use of PGD to avoid genetic disease. The same poll found that 59% of Americans approved of the use of genetic engineering to eliminate disease genes from the unborn.

That's a staggering number. It suggests that It suggests that of the four million children who will be born in the United States this year, around 2.4 million of them will be born to parents who approve of the use of genetic engineering to cure genetic diseases. Of the 120,000 American children who will be born with some sort of birth defect this year, more than 70,000 of them have parents who approve of the use of genetic engineering to correct those defects while the child is still in the womb.

These technologies - in vitro fertilization, pre-implantation genetic diagnosis, in-utero gene therapy, and genetic engineering – have all been developed in service to one thing – the human desire to have and raise a healthy child. The statistics we just discussed are a testament to how strong that desire is. IVF has already helped more than 3 million couples conceive. PGD has helped bring several thousand children into this world free of crippling genetic defects. In-utero gene therapy and genetic engineering will eventually help other couples and unborn children.

As with other medical technologies, society stands to benefit from these reproductive health techniques through a reduction in health care costs. PGD is expensive, but far less so than chemotherapy or a lifetime of care for a victim of Down Syndrome. Prevention is easier than cure. If we can use technology to reduce the incidence of disease-causing or disease-promoting genes, we all stand to benefit, and none more than the children born from these techniques themselves.

That hasn't stopped critics from attacking these techniques. The most commonly voiced objection is that meddling in reproduction is unnatural or constitutes an attempt to play God. Yet as in the other places we've heard it, that argument ignores much of human history and modern society. We've already made major changes in reproduction since our hunter-gatherer days. We use contraceptives to decouple sex from fertility. We use ultrasounds and prenatal tests to learn about our children before they're born. Our doctors perform caesarean sections to deliver difficult pregnancies. We use incubators and medicines to save the lives of sick or premature babies.

In short, we already intervene at every point we can to stack the deck in favor of having a healthy child. These new technologies are just new tools to accomplish a goal as old as our species – to bring our children safely into this world.

Chapter 8 – Designer Children

Craig Venter has been called many things - arrogant, brash, disdainful, egomaniacal, even "Darth Venter". He's seldom been called humble. The tall, intense, balding scientist has never been one to back away from a fight. It takes confidence to believe that one can challenge and beat the most ambitious scientific project of the last half century. In 1997, that's what the maverick biologist did when he founded Celera Genomics, a company devoted to sequencing the human genome in just three years. At that point, the competing public Human Genome Project (HGP) run by the National Human Genome Research Institute (NHGRI) was about 10 percent done with its own sequencing of human DNA. The HGP was funded to the tune of more than ten billion dollars and on track to finish their work in another five years. Yet with only $300 million of corporate backing and a late start, Venter set out to beat them at their own game.

Venter succeeded in sequencing the human genome in those three short years, and his challenge spurred the public HGP forward, so that the two groups simultaneously released their results in 2000, two years ahead of the public project's original schedule.

Journalist James Shreeve does a masterful job of telling the story of the personalities in the race to sequence the genome in his book *The Genome War*. Yet underlying those personal stories is a story of technology and how its progress will affect our world.

The key to Celera's success was a technique called shotgun sequencing. In normal gene sequencing, the long strands of DNA that make up a genome are first mapped out to identify key landmarks. Then the DNA is copied several times and divided into shorter overlapping segments. Those segments can then be read to determine their exact sequence. Researchers refer back to the map to determine where this particular sequence fits in the grand scheme. If you imagine the genome as a giant jigsaw puzzle, it's as if the researchers assemble the entire puzzle, then take each piece, carefully note where it is, and sequence it. By remembering where in the puzzle it belongs, they can then assemble the entire sequence.

Shotgun sequencing isn't so careful. It chops DNA up into short overlapping strands, but doesn't take any note of a map or where those strands fit. Instead, after sequencing each strand, a computer uses the overlaps between them to fit them all back together. The researchers never put the jigsaw puzzle together in the first place, which saves them time and money. While the Human Genome Project avoided shotgun sequencing as too radical and too dependent on computers, Venter made a bet on it, and the bet payed off – Celera was able to sequence the human genome at somewhere between thirty and fifty times less cost than the public project, and substantially faster as well.

Shotgun sequencing was not the first speed and cost improvement in sequencing technology. Nor was it the last. According to the NHGRI, the cost of sequencing a single base-pair, or single 'letter' of DNA, dropped by more than a factor of one hundred between 1990 and 2002, from about $10 per base pair to about 9 cents.

NHGRI's report on the topic states that "Starting in 1990 ... the reduction in sequencing costs has followed a straight line on a log curve". A pattern that follows a straight line on a log curve is one that's growing (or shrinking) exponentially. The amount of DNA you can sequence for a given price seems to be going up by about 50 percent each year, or a little more than doubling every two years.

Additional data going back all the way to the sequencing of RNA in 1965 seems to confirm the general trend, and the experts believe that it will continue. The NHGRI report states that "a two-log [factor of 100] decrease in cost in the next five years" is possible and that "there is no physical reason that we can't achieve a $1,000 genome". At the current rate of progress we'll be able to sequence the entire genome of a human being from scratch for $1,000 in the year 2035. Six years later, in 2041, we'll be able to sequence an entire human genome for just $100.

This trend has caught the attention of commentators, scientists, and entrepreneurs alike. Computer scientist Raymond Kurzweil and zoologist Richard Dawkins have both written about this exponential improvement, and connected it to the other famous exponential technology improvement of our time - Moore's Law. Moore's Law, coined by Intel founder Gordon Moore, is the observation that the number of circuits that can be placed on a computer chip - and thus the chip's processing power - is growing exponentially, doubling every eighteen months or two years, depending upon who you ask.

As it turns out, Moore's Law isn't just an analogy for our rapidly increasing gene sequencing power, it's also the *basis* of much of that power. The newest techniques for reading genes are actually built with the same technologies used to build the CPU and memory chips inside your computer.

Take gene chips as an example. Gene chips are manufactured using processes taken from the computer industry. Each gene chip is a grid with anywhere from a few thousand to more than a million strands of DNA anchored to its surface. Each tiny square on the grid carries a different DNA sequence. Those DNA strands act like genetic Velcro. When they come in contact with a matching strand, they adhere to it, forming the famous double helix. That allows researchers to identify the genes in a sample by seeing which points on the chip they stick to.

The power of a gene chip is directly related to how fine the grid is. The more distinct DNA sequences you can fit on a chip, the more genes you can search for in a blood sample. As the computer industry comes up with new methods for increasing the density of electronic chips, those techniques become available for gene chips as well. In a very real sense, the power of gene chips is pegged to Moore's Law and the computer industry.

The same is true of another technique called linear DNA analysis. Inside every cell of your body, your DNA is packed into a tight tangle. When it comes time to read your genes, enzymes stretch out parts of your DNA. Every time one of your cells divides, your entire genome must be stretched out and read. The process takes about fifteen minutes.

Linear DNA analysis works similarly. Instead of trying to read short segments of DNA like a gene chip or gene sequencer, linear DNA analyzers stretch out whole chromosomes and run them through a computer chip where a laser can scan them. Linear DNA analyzers, like gene chips, are built with techniques borrowed from the computer industry.

Gene chips and linear DNA analyzers also cheat a bit. They don't actually sequence your DNA from scratch. Instead, they rely on the fact that all humans have very similar DNA. While looking at every bit of your DNA is a daunting task, looking for the smaller number of places where your DNA might differ from the rest of humanity is more manageable.

For all that we've talked about variation in human genes, the differences are actually very small. Your genome is spelled out in pairs of molecules called base-pairs. Base pairs come in one of four types, guanine, cytosine, thymine, or adenine, commonly abbreviated as G, C, T, and A. You have about 3 billion of these genetic letters, yet only about one in one thousand base pairs is actually different from person to person. These 3 million places where humans differ are called single nucleotide polymorphisms, or SNPs. (A nucleotide is another name for a base-pair, and 'polymorphism' implies a place where the nucleotides come in multiple forms or versions.)

The fact that the DNA we humans carry differs in only 3 millions places out of 3 billion makes the task of genetic screening much simpler. Instead of having to read every letter one at a time, researchers can look for whole stretches that are known to come in just one of a few versions. For example, one stretch of DNA may appear in humans in only two versions, one that reads TTGC***A***CG and another that reads TTGC***G***CG.

By the time you read this book, leading gene-chip company Affymetrix will be selling gene-chips that detect at least one hundred thousand of these SNPs in the span of a few hours. That's still a small chunk of the 3 million SNPs that humans possess, but it's a huge step up in the number of genes that can be identified in a single test. Analyzing one hundred thousand SNPs by sequencing your genes from scratch would cost several million dollars today, while one of Affymetrix's chips costs just a few thousand dollars.

US Genomics, a startup leading the charge towards linear DNA analysis claims that its system can already scan 30 million base pairs a second for the single nucleotide polymorphisms that distinguish individuals. They estimate that in the next ten years they'll be able to map an entire genome in a day.

Affymetrix and US Genomics are just two companies in the race to provide faster, cheaper genetic screening technologies. If they fail, their competitors may yet succeed. Driven by the huge potential market, dozens of companies are working to deliver better ways to read the genome. Collectively they're fueling the exponential increase in gene sequencing and gene mapping power.

If sequencing power for a given price really is following this Moore's Law-like progression, than the future of sequencing will look something like the future of computing. Today we have cell phones, digital watches, and calculators with more computing power than the supercomputers of a few decades ago. In a few decades we may have portable lab sequencers with more sequencing power than the entire Human Genome Project.

Cheap gene sequencing would have two immediate consequences. First, the cheaper and easier it is to sequence a genome, the easier it is to determine which genes influence which mental and physical traits. Researchers try to do that today with population genetic studies. They take a number of people with a certain rare disease, for instance, and try to determine what genes they have in common with each other but not with the general population. Currently those studies are long and arduous. The tools researchers have for looking at genetic samples are fairly crude and slow. It can take years of research to identify a single gene.

In contrast, some of the techniques we just talked about could map a person's entire genome in a matter of hours within the next decade or two. That gain in speed will accelerate research into the function of our genes. It'll give us a more comprehensive view of the genes involved in heart disease, cancer, and other ailments.

Cheap gene sequencing won't stop at decoding the genes responsible for diseases, though. The technology will give researchers the tools they need to identify genes that influence physical traits like height, facial structure, musculature, eye, skin, and hair color. The same tools will help identify genes involved in mental traits too - those that impact our intelligence, mood, and personality.

That leads us to the second consequence. Cheap gene sequencing will make genetic testing of the unborn easier, faster, and more accurate. Where today's prenatal tests can identify a few dozen different disease genes, parents in the near future will be able to identify every one of the thirty thousand or so genes their unborn child carries. Combined with knowledge of what these genes do, that will give parents an idea of a prospective child's appearance, intelligence, and personality.

With knowledge comes power. A family going through IVF, for instance, could use PGD to completely screen the genes of the embryos they've created and select a few for implantation based on the traits they find.

Princeton's Lee Silver, in his 1997 book *Remaking Eden*, paints an evocative picture of how this might work. A couple goes into an infertility lab and goes through IVF. The IVF procedure produces several embryos, each of which has its genes examined via PGD. Now the couple sits in front of a computer screen. Displayed on the screen are genetic "profiles" of the embryos they've created, and computer generated pictures showing the children and adults those embryos are likely to grow up to be.

The parents click on one of the profiles and see the face of a potential child at age sixteen. After all, the shape of a nose, the color of eyes, the tone of skin, and much more are encoded in our genes. Each profile also shows any severe genetic diseases the embryo carries, how high the embryo's risk is of developing more complex diseases like heart disease and cancer, and even an assessment of the child's likely personality. Click - a lovely green-eyed, brown-haired girl, but with a high risk of heart disease and only average intelligence. Click – a pudgy, probably shy boy likely to have a high IQ. Click – a tall, athletic girl with perfect pitch but a tendency towards manic depression. Click – another possible child. Click – yet another.

Yet PGD is inherently limited. Parents can only choose from the set of embryos created in a single round of IVF. This can be anywhere from one to twenty, but seldom more. With only a few embryos to choose from, there will be plusses and minuses to each. One embryo might carry genes for great looks and high potential intelligence, but also a propensity for anxiety. Another might carry a propensity for a sunny disposition and perfect pitch but a propensity for obesity.

Doubtlessly, some parents will draw the line at merely peeking at their future children's genes. Others, those going through IVF, will need to choose which embryos to implant anyway. At least some of them will be interested in making that decision based on these other factors that genetic tests reveal, but will be content in simply selecting one of the available embryos without any attempt to alter its genes.

Yet as we discussed last chapter, there will be scenarios in which parents will have good medical reasons to alter the genes of their unborn children. Once those medical uses of genetic engineering or in-utero gene therapy are accepted, it may be a short step to altering genes for more cosmetic or enhancement reasons.

Westerners are not very friendly to the idea of using genetic technology to enhance the unborn. Compared to the 73 percent of Americans who would favor using PGD to screen out disease, only 22 percent approve of using PGD to select desirable features. Compared to 59 percent of Americans who approve of using genetic engineering to remove genetic diseases, only 20 percent approve of using the technology to create desirable features in the unborn.

As we saw a few chapters ago, support for genetic technologies is much higher in other parts of the world, especially Asia, where majorities in several countries approve of the use of genetic engineering to select traits of children.

It's possible that attitudes will change in western nations as well. According to the Johns Hopkins University Genetics and Public Policy Center, only 10 percent of Americans approved of the use of genetic engineering to create desired traits in children in 1994, compared to 20 percent in 2002. What's more, the attitudes are related to age. 14 percent of those in the 50-and-older age group are highly supportive of genetic technologies (as measured by the Hopkins study) vs. 27 percent of those in the 18-29 age group.

There's also a large gray area between the widely supported medical uses of genetic technologies and the less supported enhancement uses. Preventing Down Syndrome is clearly a medical use. Screening out or using genetic engineering to remove a gene that increases the likelihood of some disease is also medical. What about selecting for or engineering in a gene that dramatically reduces the risk of heart disease or cancer to far below the population norm?

For example, we've already discussed the ApoE e4 allele, which dramatically increases the risk of both Alzheimer's disease and high cholesterol. A different variant of ApoE - the e2 allele - reduces risk below the population norm. Some might consider splicing in that gene an enhancement, while others considered it a medical use.

Or consider the CCR-5 gene. About 1 percent of the population carry a variant of this gene that makes them immune to most strains of HIV. While altering a child's genome to carry this variant would have no known effect other than resistance to a disease (AIDS), it could still be viewed as an enhancement, as it conveys a trait that few members of the population have today.

As a final case, consider obesity. There are at least forty different mutations of six human genes that are thought to increase the likelihood of obesity. Obesity itself is a risk factor for heart disease, diabetes, and cancer. Reducing the risk of obesity in a child clearly has medical value. Just as clearly it has a cosmetic value. On medical grounds society might allow genetic engineering to reduce the risk of obesity. Yet there's no way to know if parents are choosing such a gene because they want their child to be healthy or because they want their child to be attractive. No matter what the motive, engineering a child's future weight is certainly cosmetic. If it's allowed, then people will start to wonder why they can't genetically engineer height, or hair color, or skin tone.

There are dozens of other examples like these, and science will only discover more as time goes on. These gray areas form the slippery slope from therapy to enhancement. Once you accept the idea of genetically altering an embryo to remove a gene that might cause cancer, it's not a far step to adding a gene that reduces the risk of cancer. Once you accept adding a gene that reduces the risk of a disease, a gene that increases longevity and thus reduces the risk of many diseases seems more reasonable. Once society is comfortable with splicing in genes to reduce a child's risk of obesity (and thus stave off heart disease and diabetes), choosing genes that promote good looks or intelligence doesn't seem so shocking.

History suggests that in general new biological technologies are feared at first, and then accepted as they prove their worth. IVF, organ transplants, vaccination, and other technologies were met with hostility when they first appeared but are now considered routine. What we've just seen suggests that genetic enhancements could go the same route. In Asia a majority of the population already supports the use of genetic technologies to enhance children. In the west it seems likely that the strong support for medical uses of genetic technology will gradually shift to bolster the small but already growing support for genetic enhancements.

Cloning

The most widely feared and denounced genetic technology is reproductive cloning– the production of a human baby with genes copied from another person. Researchers have cloned sheep, cows, and monkeys, and dubious claims have been made that humans have been cloned. Whether that's happened yet or not, the fundamental technology exists, and the claims have roused public fears.

Cloning has been almost uniformly denounced as morally repugnant. Leon Kass, chairman of the President's Council on Bioethics (PCBE), describes human cloning as a threat to "the dignity of human procreation" and a "first step toward a eugenic world in which children become objects of manipulation and products of will." Virtually every political leader in every country is publicly opposed to cloned babies. More than 100 countries at the UN support a ban on reproductive cloning, though the ban has not yet passed due to disagreement over whether *therapeutic cloning*, which produces medically useful stem cells but no children, should also be banned.

There are certain cases where reproductive cloning is the best or only way to have a biologically related child. For example, consider a couple in which neither the man's sperm nor the woman's eggs are viable. Neither IVF nor egg donors nor sperm donors nor surrogate mothers could help this couple. Cloning, on the other hand, could give them a child that shared the genes of one of the parents.

Same-sex couples are another case where cloning makes biological sense. Without cloning a lesbian couple could use a sperm donor to conceive, and a gay male couple could find a willing surrogate mom and egg donor. Both of these solutions, though, result in a child with only half of the genes of one of the parents in the couple. And both solutions involve a third party who's now biologically related to the child and may produce unwanted entanglements. A clone, on the other hand, is related only to one member of the couple. Especially for lesbian couples (who don't need the services of a surrogate mother), this may be the best option for having a biologically related child.

Reproductive cloning isn't safe today. Cloned animals are often born with deformities or health problems. So long as that's the case, it's unethical to attempt it with a human. The technique can and should be illegal until basic research on animals makes us confident that it will work safely in humans. However, the anti-cloning movement isn't primarily concerned with safety. It's concerned with humanity and identity, and the notion that giving birth to cloned humans would somehow rob them - or the rest of us - of some unique human dignity.

The debate about reproductive cloning is often marred by a misunderstanding of what it means to clone someone. In popular parlance, clone is shorthand for "copy". Imagine a clone of yourself. What do you see?

You may be picturing someone about your age, who looks just like you, with your same beliefs and attitudes, perhaps even your same skills and knowledge of the world. But this would not be your clone.

In fact, there are 40 million human clones alive today. We call them identical twins. We recognize them as individuals without a second thought. We have no angst about whether a twin is less of a person than a singleton. We have no doubt that each twin has a soul, at least in so far as the rest of us have souls.

Clones conceived with the help of biotechnology would be just as much individuals as are naturally conceived identical twins. In fact, a clone of you would differ from you much more than a twin. Your clone would grow up decades later than you did, surrounded by different people and the culture of a different era. He or she would have a personality and beliefs different from yours. There is no need to ban reproductive cloning on moral grounds, because clones would be ordinary people like the rest of us.

Genes and Human Behavior

One of the major concerns among those who oppose using genetic technology to select desirable traits of the unborn is the amount of power this could give parents over their children. Environmentalist Bill McKibben is among those who's concerned about this. In his book *Enough: Staying Human in an Engineered Age*, he writes:

> "The person left without any choice *at all* is the one you've engineered. You've decided, for once and for all, certain things about him; ... You'll be taking this on your own shoulders, exercising infinitely more power over your child than your parents did over you. ... He may have no more choice about how to live his life than a Hindu born untouchable."

How realistic is McKibben's concern? The answer to that question depends on many factors, and particularly on just how much our genes affect who we are.

It's possible to use genetic engineering to alter the odds that a child will have any given trait. Exactly how strong an effect you can have on the odds depends on what you're trying to change. Some traits, like eye color, are almost completely genetic. Other traits, like height, are largely genetic but also depend upon a proper environment - in this case one with adequate nutrition. Most traits, however, depend upon heavily on both genes and environment. A child with the muscle-affecting genes of Arnold Schwarzenegger won't necessarily grow up to have a massive physique. Those muscles require both the right genes and long hours spent in the gym.

Behavioral traits are even more complex. They depend on a complex interaction of genes and environment. Some genes affect personality only in the presence of some environmental triggers. For example, in 2002 Avshalom Caspi and collaborators at Kings College in London concluded that children with a mutation in the gene for the brain enzyme monoamine oxidase A (MAOA) are more likely to commit acts of violence, but *only* if they're abused while young. Children with the MAOA mutation who grow up in a supportive, non-abusive home are no more likely to be violent than children without the mutation. Abused children who *don't* have the MAOA mutation are also unlikely to grow up to become violent. Neither the environment nor the gene produces an effect in isolation. They *both* have to be present to affect behavior.

Despite this complexity, we can make some general observations of the degree to which genes affect various personality traits. These observations come from studies of families - particularly adopted children, twins raised in the same household, and twins separated at birth.

These studies work by comparing the scores of various pairs of siblings and sometimes parents on a variety of tests. What they show is that genes have a substantial impact on personality, but not a dominant one.

For example, imagine two pairs of twins. One pair is made up of identical twins. Those twins share the same genome. The second pair is made up of fraternal twins of the same gender. The fraternal twins only share half their genes with one another. So in a sense, the identical twins are twice as genetically related.

When researchers administer tests to these pairs, they find that the identical twins tend to have very similar scores. The similarity between their scores is *greater than the similarity between the scores of the fraternal twins.* Since both pairs grew up together in similar environments, greater genetic similarity is likely the cause of this greater similarity in scores among the identical twins.

The same things can be seen in many other cases. Identical twins separated at birth tend to have similar scores on a variety of personality and intelligence tests. On most of these tests, adopted children tend to resemble their biological parents more than their adopted parents. Overall the data suggests that genes play a large role in intelligence and personality.

On the other hand, genes don't control everything. Identical twins are still quite different from one another. Identical twins raised apart from each other are just as varied as unrelated people of the same age in rates of smoking or church attendance or numerous other factors. And even on tests of intelligence and personality, identical twins separated at birth have wildly varying scores - they're just more similar to each other than the scores of unrelated people or fraternal twins.

So how large is the effect of genes on those traits that they do influence? Let's consider how they score on tests of intelligence and personality.

The word 'intelligence' means many things to many people. So researchers have instead focused on IQ, or on a factor called 'g'. Tests of IQ look at simple problem solving abilities, verbal skills, and spatial skills. They doesn't measure leadership abilities, depth of knowledge, motivation, creativity, emotional intelligence, interpersonal skills, or any of dozens of other things you may think of when you call someone intelligent.

Despite that, IQ is a relevant number. While different IQ tests differ in whether they're timed, whether they focus on verbal or spatial skills, how long they are, or many other factors, individuals tend to get fairly consistent scores on these tests. What's more, according to a wide variety of studies,

IQ is a fairly good predictor of educational achievement, of income later in life, and even in some cases of future life expectancy. So if IQ tests don't really measure the full gamut of 'intelligence' they're at least measuring some things that we care about.

Researchers express the power of genes on a trait in terms of correlations. Correlations are on a scale of 0 to 1. A correlation of 0 means the traits are completely unrelated. A correlation of 1 means that one trait perfectly predicts another.

The correlation between genes and IQ is somewhere around 0.5. Most estimates have ranged from 0.35 to 0.75. This indicates that genes and environment play a fairly equal role in determining IQ. Which in turn means that the people with the highest IQ tend to be those with *both* the genes that encourage high IQ and an environment that encouraged high IQ.

Personality is, if anything, even more difficult to quantify than intelligence. The most agreed upon model of psychologists is to divide personality traits into the so-called "Big Five": Neuroticism, Introversion-Extroversion, Agreeableness, Conscientiousness, and Openness to Experience.

Like attempts to quantify intelligence, attempts to quantify personality in these traits are controversial. There are several competing models, but the "Big Five" model is the most popular among psychologists of personality.

Based on twin studies, researchers estimate the genetic contribution to these traits as somewhere between 0.3 and 0.5. Research from adoption studies produces a lower range of estimates, between about 0.2 and 0.35. So in general it appears that genetic variations account for about half of variations in IQ, and between a half and a third of variations in personality test scores.

This one-third-to-one-half correlation between genes and behavioral traits constrains the amount of power that genetic engineering will give parents over their children. It indicates that genetic engineering will give parents a tool that can increase their odds of having a certain kind of child, but that no amount of genetic tinkering will allow parents to completely control the IQ or personality of a child. The final personality, intelligence, and other behavioral traits of your child will still depend critically on environment and random chance.

To illustrate this, let's do a quick thought experiment. Suppose that you want to have a high IQ child. You could increase the odds of this by giving your child the same genes as someone who's already lived and who had a high IQ. For argument's sake, let's say you pick Albert Einstein. Since this is a purely hypothetical case, let's also ignore all the issues involved in discovering his genetic makeup, determining which genes are involved in intelligence, and actually performing the genetic engineering.

We don't know what Einstein's IQ was, but it's been guessed at about 160. If Einstein did have a 160 IQ, and you conceive a child with the same intelligence-linked genes as Einstein, what's your child's IQ most likely to be? You may be tempted to guess 160, but remember that only about half of the variation of IQ from the norm is accounted for by genetic variations. The other half is a mixture of environment and chance. Einstein's hypothetical IQ differed from the norm by 60 points. So we should expect that about half of this variance, or 30 points, is accounted for by his genetic differences from the human norm.

That means that if we created a population of Einstein clones, and raised them in "average" environments, their average IQ would be 30 points higher than the norm, or 130. The individuals of the group would all have different IQs, of course, accounted for by their different environments and random events in their lives. Those IQs would form a normal bell-curve shaped distribution with 130 in the center. Fewer than one out of every 400 of these clones would have an IQ of 160 or higher. Just as many would have an IQ of 100 or lower. Your odds of producing an Einstein-level genius by copying his genes are just as high as your odds of producing someone of average intelligence.

On the other hand, 99.8 percent of these children are going to be above the human average in IQ. More than 80 percent will have IQs of 120 or higher. And these numbers assume "average" environments. Copying Einstein's genes isn't likely to produce a super-genius, but it's very likely to result in a child with a substantially above- average IQ.

This illustrates both the power and limitations of genetic engineering. Children carrying genes linked to high IQ are going to, *on average*, have IQs higher than the norm. But the only ones to reach truly high IQs are going to have environments and random life events just as exceptional as their genes.

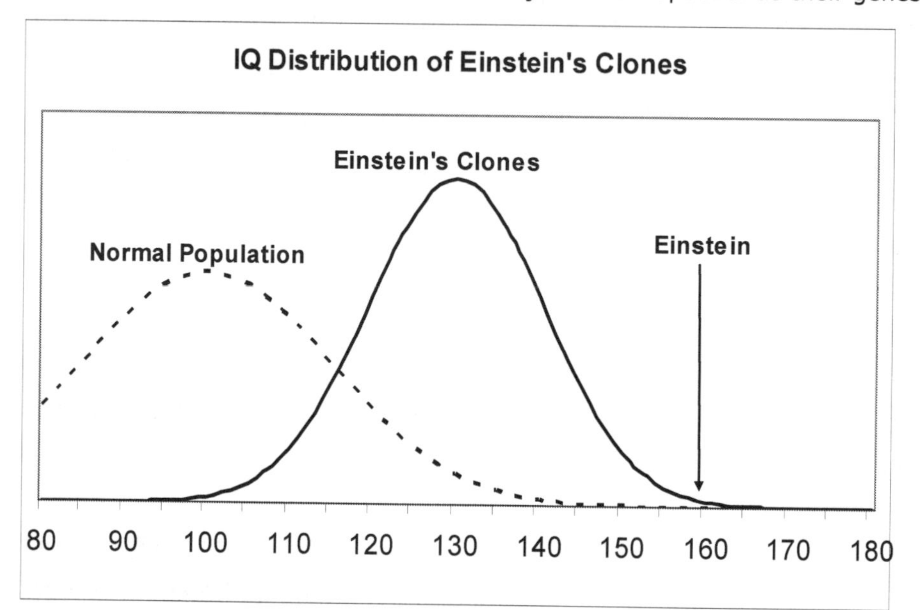

Figure 9 - Distribution of IQ. If Einstein had a 160 IQ which was 50 percent accounted for by genes and 50 percent by environment and chance, his clones would have an average IQ of 130. Only 1 in 400 of his clones would have an IQ equal to or greater than Einstein himself.

To be fair, we've played a bit fast and loose with genetics in this thought experiment. The truth is that we don't know Einstein's IQ for certain. Even if we did, we wouldn't be able to say how much his high IQ stemmed from exceptional genes and exceptional environment. The correlations between genes and IQ hold true for a *population*. They don't say anything certain about a given individual. Perhaps in Einstein's particular case his environment was more exceptional than his genetics, or vice versa. Still, the overall point holds, and we can state the thought experiment a bit more precisely. If you took a large number of individuals with 160 IQs, cloned them, and raised the children in randomly selected homes, the average IQs of the children would be 130. Exceptional individuals *usually* have both exceptional genes and exceptional environments.

The same basic truth applies to personality measures. Selecting genes associated with high a high degree of openness to experience, for example, increases the *likelihood* that you'll get a very open child, but doesn't guarantee it.

What's more, genetic alterations of personality come with another risk – that of *overshooting* the target. If you genetically engineer an embryo to select genes associated with high agreeableness, you increase the odds of getting a pleasant, agreeable child. You also increase the odds of getting a child who's agreeable to a fault. If you genetically engineer your child to increase the odds that he'll be an aggressive go-getter, you also increase the odds that he'll become an overbearing boor.

Linguist and evolutionary psychologist Stephen Pinker pointed out this problem in an essay in the *Boston Globe*.

> "... most traits are desirable at intermediate values. Wallis Simpson said that you can't be too rich or too thin, but other traits don't work that way. Take aggressiveness. Parents don't want their children to be punching bags or doormats, but they also don't want Attila the Hun either. Most want their children to face life with confidence rather than sitting at home cowering in fear, but they don't want a reckless daredevil out of Jackass."

The final limitation on genetic control of personality is the fact that any given gene plays multiple roles. A gene that affects one facet of personality or intelligence is likely to affect others. For example, variations in the COMT gene are believed to account for 4 percent of human variation in IQ. COMT comes in two varieties – 'val' and 'met'. Everyone is born with two copies of the gene. People born with two copies of the 'met' version of the gene seem to make more efficient use of the frontal lobes of their brain. They tend to have better working memories and perform better than other people on various tests. Yet they also seem to be more prone to anxiety, more sensitive to pain, and more likely to develop schizophrenia than people with two copies of the 'val' version, or one each of 'val' and 'met'.

This is just one example. As we start to decipher the role of genes in our behavior, we'll find that many - if not all - play multiple roles. That means that there will always be trade offs. Perhaps a parent will have to decide if that increased likelihood of high IQ is worth the increased chance of anxiety or mental disorder. There will be thousands of other genes that come with such trade-offs.

These trade offs and other limitations shed light on another concern about genetic engineering: the worry that power over our children's genes will homogenize society. Few parents with access to genetic technology are going to choose to have children with bubble boy disease or muscular dystrophy, and other traits like propensity for heart disease and cancer will probably not be popular. But among the behavioral traits, the trade offs are so numerous and control so imprecise that parents are bound to bring a wide menagerie of children into this world.

Genes tremendously impact our behavior, and in the near future parents will be able to employ genetic engineering to influence their children's health and appearance. They'll be able to dictate hair color, eye color, skin tone, height, facial structure, and much more. They'll be able to affect health factors, athleticism, and weight. And they'll be able to influence - but never completely control - their child's IQ, personality, and other mental traits. Genetic engineering will give parents substantial choice in their children's personalities, but far from absolute power.

This is not to say that cloning and genetic engineering won't cause any problems. In 2001 Margret Talbot wrote an article on cloning in *The New York Times* magazine that illustrates one realistic concern. Talbot talked to families who had lost a child and were interested in cloning their deceased son or daughter. The families all claimed to understand that two children who shared the same genes would be unique individuals - that nothing could bring their son or daughter back to life. But despite this they seemed to cling to an emotional view of cloning as a way to resurrect their lost child.

A clone will never be the same as the dead child he or she is cloned from. Parents who think they're bringing back their lost son or daughter are bound to be disappointed, and a child always measured against an invisible yardstick of someone else's life is going to suffer.

We can anticipate such problems arising with genetic engineering as well. A failed athlete may engineer his son for physical abilities, and push that child into sports whether the child likes it or not. If the child fails to excel, the father's reaction may range from severe disappointment to emotional or even physical abuse. An academic may engineer her daughter for extreme intelligence but react in dismay if she chooses a career in art rather than science.

Even with good parenting, the emotional consequences for a child who has been "engineered" might be negative and severe. Children who know that they were "designed" to have a certain trait may feel deeply resentful. They may feel as if their lives have been controlled and scripted from the outset, leaving them no chance for individuality. They may even resent their parents for not engineering them *well enough* - providing them with great athletic abilities when what they really want is to attend Harvard, or not being willing to sacrifice financially to buy them better looks or a longer lifespan.

This is not a new concern. When the first babies conceived via in vitro fertilization were born, many critics of the technique claimed that these children would be emotionally scarred for life, either by the process of gestating for a few days in a test tube or, more likely, by the very knowledge that they are 'artificial' children, born in an 'unnatural' way. Biotech gadfly Jeremy Rifkin warned about the "psychological implications of growing up sheltered not by a warm womb but by steel and glass, belonging to no one but the lab technician who joined together sperm and egg". Even as recently as 2002, Leon Kass warned that any conception that occurs in a laboratory is "profoundly dehumanizing, no matter how genetically good or healthy the resultant children".

Other bioethicists warned of other psychological problems for IVF children. Some pointed to the stress of past failures to conceive or of the IVF process itself as a reason why IVF families might be weaker. Others suggested that an infertile couple who placed all their dreams in IVF might conjure up a 'fantasy child' that no real child could live up to.

Yet psychological studies years later show that IVF children have grown up to be perfectly normal and well adjusted. In fact, if anything, they tend to be *more* emotionally healthy than the norm. The data suggests that parents who went to the effort to use in vitro fertilization provided a healthier upbringing for their kids, on average, than parents who conceived naturally.

In a 1998 review published in the journal *Biomedical Ethics*, researchers Emma Goodman and Susan Golombok summarized data from all the studies on IVF families published to date and concluded that "the children themselves seem to be developing normally" and that in IVF families "the quality of parenting appears to be superior, with IVF mothers displaying higher levels of warmth and emotional involvement, and both parents interacting more with the children than in naturally conceived families."

That may reflect the fact that parents of IVF children obviously wanted to have those kids. They committed hours or days researching the technique, visiting doctors, and ultimately spent thousands of dollars just to conceive their child. It's reasonable to expect that they'd be committed and involved parents as well.

The same logic extends to parents who genetically engineer their children, especially while the techniques are still fairly new and expensive. Parents who genetically engineer their children are going a step beyond IVF. They're molding the child, not just trying to bring him or her safely into the world, but trying to pass on what advantages they can. Parents who go to this extra trouble and expense *care*. These are the kinds of parents that are likely to be active in their children's lives and produce healthier and better off kids for it, even without the genetic advantages they'll convey.

Genetic engineering our children is a new idea, but investing in them is a long human tradition. Over the course of human history parents have always sought to give their children whatever advantages they could. Parents have tried to feed, clothe, house, and teach their children as well as they were able. Those advantages have in many cases altered the brains and bodies of those children for life.

Indeed, parental investment in children is selected for by evolution. Any parent who doesn't pour energy into giving his or her children the best possible place in life is going to be at a disadvantage in passing genes on to the next generation. Unlike other animals that can have dozens or even thousands of offspring in a generation, we humans have only a few, and they take longer than the children of any other species to grow to maturity. As a result, we're even more biologically engineered to care for, nurture, and try to advantage our children than parents of other species.

Our societies welcome this, celebrate it, and build institutions around it. Our schools, vaccination programs, piano lessons, private tutors, and more are all attempts to pass on whatever benefits we can to our children. As bioethicists Arthur Caplan notes:

> "The notion of enhancement is deeply ingrained in our culture. The only debate about the Kaplan SAT prep test is how many weeks to attend. What about spending $25,000 a year to send your kid to private school? Nobody says, `You should be ashamed, you're giving your child an advantage.'"

Schooling, nutrition, and other cultural phenomena can make a lasting difference. For example, better nutrition is credited for increasing average height of men and women in the United States by around 3″ in the last two centuries, from 5′ 6″ and 5′ ½″ respectively in the colonial period, to 5′ 9″ and 5′ 3 ½ " today. Similar changes have occurred in other countries around the world.

In 1984, political scientist James Flynn noticed another interesting trend, this time in human IQ. IQ scores are normalized to one hundred - every year, no matter how well people perform, the average performance is called one hundred, and the other scores are arranged on a curve around it. For that reason, it's easy to miss something deeper going on. Flynn took the *raw* score on IQ tests – the number of questions answered correctly on incorrectly – and found that it had been increasingly steadily over time. In the US, Flynn saw a twenty-five point gain in IQ tests between 1918 and 1995. That is to say, if the tests of people who scored one hundred in 1918 were scored against the 1995 curve, they'd result in an IQ score of seventy-five. If the 1995 tests that scored one hundred were evaluated against the 1918 curve, they'd score at one hundred and twenty-five.

Flynn went on to find the same effect in multiple intelligence tests from different countries over different periods of the twentieth century. The test results from the United States were on an IQ test called the Stanford-Binet. In West Germany IQs rose by twenty points between 1954 and 1981 on a different test called the WISC. In France they rose between twenty to 25 points between 1949 and 1974, on a test called Raven's Progressive Matrices.

Interestingly enough, the best documented gains are not on tests which measure crystallized IQ, or concrete knowledge of vocabulary and facts. Instead they're on tests of fluid IQ, which measure abstract reasoning, spatial skills, and visual skills. Across the world the average rise has been around 3 points of raw score per decade, with results to back this from Japan, the Netherlands, the UK, Greece, and dozens of other countries.

The Flynn effect can't be put down to genetics because of the incredible speed of the effect. Flynn went on to show this by documenting a large increase even between fathers and their own sons. There's active debate over exactly what causes the effect, but most speculation centers around better nutrition, better schooling, and increased exposure to information of all sorts during childhood.

In both these cases – height and IQ – children of past generations have been enhanced through non-genetic techniques. Yet at the end, the effect is as real. Whether a child grows up to be taller or more mentally nimble as a result of genetic changes or environmental ones doesn't really matter to the child or the parent. What does matter, to almost every parent, is to do the best we can for our children, affording them the advantages we can, as nature has engineered us to do.

For both cultural and biological reasons, it's a basic tenet of society today that parents are the best people to make choices on behalf of their children, until the children are old enough to choose for themselves, or unless the parents demonstrate obviously destructive behavior. We trust parents to choose the schools children go to, the church they attend, the language they learn, the food they eat, the television and movies they watch, and more. This trust reflects the fact that, for the most part, parents want what's best for their children, and are in a better position to look after the best interests of their offspring than anyone else.

It's interesting, then, that some opponents of genetic enhancement have labeled the practice *eugenic*. For example, Presidential Council on Bio-Ethics Chairman Leon Kass, who has made it clear that he would like to see a ban on genetic engineering, has written that the debate over cloning offers a unique practical opportunity to combat "the impending practice of germline genetic manipulation or other eugenic practices." Boston University bioethicists George Annas, who's proposed a UN ban on genetic engineering, went so far as to say that "Modern genetics is eugenics," while visiting the Holocaust Museum in Washington D.C.

And Gilbert Meilander, another member of the PCBE, wrote in 2001 in an essay in *First Things* magazine that:

> "our present condition is this: we have entered a new era of eugenics. That science which attempts to improve the inherited characteristics of the species and which had gone so suddenly out of fashion after World War II and the Nazi doctors now climbs steadily back toward respectability"

The connection to WWII Germany here is quite explicit. While 'eugenics' may at one point have referred to efforts to improve mankind, the association today is with attempts by totalitarian states to control the makeup of the population, usually at the barrel of a gun. The use of the word is intended to form an association between any attempt to influence the genes of one's child and WWII concentration camps, gas chambers, and Nazi human experiments.

What's interesting is that in the debate over human enhancement, the only people advocating state control over the genetic makeup of the population are those who would like to see genetic enhancement techniques *prohibited*. The advocates of human enhancement, on the other hand, are arguing for individual and family *choice*, the opposite of state control.

To a certain extent then, the prohibitionists are on the eugenic side of this debate. It's those who oppose individual and family genetic choice who have, in essence, decided that there's a certain "correct" genetic heritage for humanity (the one we have today) and that the populace should not be allowed any choice in the matter. The relatively small number of advocates for genetic choice, on the other hand, aren't trying to impose their opinion on the rest of the country or the rest of the world. They're only seeking the freedom to make genetic decisions for themselves and their families - the same freedom they enjoy to make the thousand other decisions involved in raising a child.

Regulating Reproductive Technologies

A regulatory regime consistent with family choice would focus on safety, education, and equality rather than prohibition. Today, regulation of reproductive technologies varies by country and by the specifics of the technology. The most consistent international regulation is around reproductive cloning. More than 30 countries, including Russia, Japan, China, and most of Europe have banned reproductive cloning. In 2003, efforts to ban cloning in the United States and internationally through a UN agreement both failed, despite widespread support for a ban on *reproductive* cloning, due to disagreement over whether bans should also cover therapeutic cloning for the production of stem cells. Nevertheless, reproductive cloning will likely be illegal throughout the entire world - or close to it - within a few years.

At the other end of the spectrum is in-vitro fertilization. IVF is legal in most countries around the world. In some countries (Japan, Australia, part of Europe) the cost of the procedure is subsidized by the government.

Neither genetic engineering of human embryos nor in-utero gene therapy have ever been attempted (so far is we know) and so their legality in most countries has not been tested. A few countries such as Germany already have laws on the books that ban such procedures, but most do not.

The most interesting case - and perhaps the best view into the future - is pre-implantation genetic diagnosis (PGD). PGD laws differ quite a bit from country to country. In the United States, PGD is legal and is virtually unregulated. Professional standards for physicians mandate follow-up studies, but there's no regulatory agency whose approval is required before performing a procedure. Nor is there any agency whose approval is required prior to the introduction of entirely new medical technique aimed at improving reproduction.

PGD is also legal in China and India, though both countries outlaw its use as a means to select the gender of a child. In Japan the technique falls outside of the guidelines of the Obstetrics and Gynecology society, but isn't prohibited by law. At least one doctor is already providing PGD as a service. In Australia the technique is legal in most states but with slightly different criteria for when it may or may not be used. In Europe laws range from fairly permissive (Belgium) to complete prohibition (Germany). In England, where PGD was first performed, every couple seeking to use the technique (or any new reproductive technique) must first apply to the Human Fertilisation and Embryology Authority. The HFEA approves requests whose goal is produce a child for an infertile couple or to avoid passing a clear genetic disease down to a descendent, and rejects almost all others.

The inconsistency of laws between different countries reflects local cultural values but has international consequences. It's clear that already parents are crossing national borders to seek out the technique. For example, a 2003 study indicated that roughly two thirds of the couples receiving PGD services in Belgium, which has Europe's most permissive PGD laws, come from other countries. There have also been highly publicized cases of British couples who, after having been turned down in their request for PGD at home, traveled to the United States for the same service.

The implication here is that unless laws covering genetic enhancements are consistent between countries, they won't stop motivated parents. Given what we've seen of attitudes towards genetic technologies in different countries, it's unlikely that the laws will ever be fully consistent. As we observed earlier, adding the burden of travel also increases the cost of a service, tilting access away from the poor and towards the rich.

This is not to say that the field should be unregulated. Regulation ought to focus on safety, education, and equality. In this respect, the United States in particular needs more regulation than it has. At a national level there's no law or regulatory agency that serves as a gatekeeper for new reproductive techniques or evaluates their safety before they're attempted. It's fortunate that both IVF and PGD (which were developed under similar conditions in England) have turned out to be quite safe and have produced healthy children, but there's no legal or scientific review process in place to ensure the safety of the next innovation in reproductive medicine. Prior to their first use in humans, reproductive techniques ought to go through the same sort of pre-clinical trials the US Food and Drug Administration requires for drugs and gene therapies.

At the point at which reproductive techniques *are* shown to be safe, the role of government should not be to deny families access but to promote education and equality. Education in this area is particularly challenging due to the common misperceptions about clones and genetic engineering. Any educational campaign has to focus on communicating the fact that every child is an individual, and that genetic techniques cannot guarantee a particular personality, intelligence, degree of health, or even appearance.

As a practical matter, a regulatory model grounded in individual and family choice to employ genetic technologies isn't likely to be adopted anytime soon. Reproductive cloning, in particular, is likely to become illegal throughout the world and remain that way for the foreseeable future. The flashpoints of debate for the next decade will be the techniques that are regulated inconsistently or not at all today - PGD, in-utero gene therapy, and genetic engineering.

To a certain extent, so long as these techniques remain legal in at least one major country, the debate will remain alive. In particular, support for the medical uses of genetic technologies is high in the United States and Asia, and the precedent of Europeans going abroad to seek out reproductive services has been set. And over time, forays into the gray area between therapy and enhancement may acclimate the public to the use of genetic technologies to alter a wider set of traits.

ULTIMATELY, WHATEVER CHOICES we may make for our children will be subject to change, at their choice, when they reach adulthood. As the other chapters of this book detail, in the coming years, pharmaceuticals, adult gene therapy, and the integration of computers into the brain will give people far more control over their own minds and bodies than what we have today.

Genetic engineering will give us the power to influence the *initial nature* of our children, but that nature will become more and more malleable over time, as new technologies that allow our children to modify their adult minds and bodies mature. In adulthood our children will be able to choose who *they* are, regardless of how we conceive and raise them.

Chapter 9 – The Wired Brain

Johnny Ray was on the phone when the stroke hit. In November of 1997, the 53 year old Vietnam veteran had seemed perfectly healthy. He spent his days as a drywall contractor and his evenings and weekends playing blues guitar. He had no reason to expect a major health problem, yet everything in his life was about to change.

Ray woke up in a different world after his stroke. Weeks had passed. He was no longer in his home in Douglassville, Georgia. Instead he was in a hospital bed in Atlanta, at the local Veterans Affairs Medical Center. From the neck down, he was completely paralyzed. A tracheotomy had taken away his voice. He could hear, think and understand, but he could not act. He could only express himself by blinking his eyelids - once for no, twice for yes.

In the US, 2,000 people suffer strokes *every day*. More than 700,000 have one each year. Other problems like ALS and neck injuries take their own toll on the ability of men and women to move their limbs and control their bodies. There are at least a quarter million quadriplegics in the United States. Worldwide there are half a million "locked in" patients like Johnny Ray who suffer more than just neck-down paralysis. They're unable to eat, drink, or talk on their own. They can observe the world but not interact with it. For them, virtually any ability to express themselves would be an immense improvement in their quality of life.

Just across Atlanta from the Veterans Affairs Medical Center is Emory University. At Emory, a neurologist named Phil Kennedy had been working on an implantable device to restore some degree of freedom to locked in patients like Johnny Ray. Over the years at Emory, his research with monkeys had shown that an electrode implanted in the brain could facilitate some kind of communication to the outside world, but no one had ever attempted such a procedure on a human being.

In person Kennedy comes across as a tall, patient Irish gent who's been transplanted into the American south. He's modest and careful with his words, while radiating a kind of solid determination. In this case, he was determined to see his technology help patients like Johnny Ray. Based on his animal studies, Kennedy was able to make a persuasive case to the FDA for a human trial. Crucial in his proposal was that the electrodes would be *wireless*. After the surgery was complete, there would be no wire protruding through the skull into the brain, limiting the risk of infection. And as for the other risks of the surgery – the patients were willing, they had very little left to lose, and a tremendous amount to gain. The FDA approved.

In March of 1998, neurosurgeon Roy Bakay implanted the device into Ray's motor cortex – the part of the brain that deals with motion. Specifically he implanted it into the area of the motor cortex responsible for moving the left hand. The cortical control system, as Kennedy calls his implant, had just a single electrode. It seemed almost impossible that it could pull out a useful signal from the hundreds of millions or even *billions* of neurons involved in motor control.

Nevertheless, animal studies had suggested it was possible. Johnny Ray was put through an intensive training program. He would stare at a computer monitor with an on-screen keyboard and a thought-controlled cursor with which to pick out letters. Under Kennedy's guidance, Ray would think about moving his left hand – up if he wanted the cursor to move up, down if he wanted the cursor to move down, and so on. As he imagined moving his hand, the electrode in his brain would pick up the signals of the few neurons near it and broadcast those to a nearby computer. The computer, in turn, would move the cursor. Something amazing was happening. A human was just *thinking* about something and a computer was responding, the stuff of science fiction.

Learning to move the cursor wasn't easy, though. The electronic pointer didn't always go where Johnny wanted it. Gaining control over it was like learning to move his hand all over again. It was an exhausting trial and error process. Yet little by little, Johnny got the hang of it. Several months later he was typing out names and sentences, and clicking icons to say things like "I'm hungry". What's more, by Johnny's reports, he had *stopped thinking about moving his hand.* He simply focused on the letters and icons he wanted to click. He willed the cursor to move without any kind of intermediary. The computer, in some sense, had become a part of him.

Today there are millions of people worldwide who lack some basic human ability that could be restored by a brain-computer interface — blind people who could see again, deaf people who could hear again, paralytics like Johnny Ray who could communicate and move again, and those with severe brain damage who could think again.

Researchers have already made dramatic progress in assisting patients with these problems. Over the last few years, experiments have restored sight to the blind and hearing to the completely deaf. Brain implants have been used to control the tremors of Parkinson's disease and the seizures of epilepsy. Animal studies have produced the sensations of touch, pleasure, and pain. Current work is looking at restoring damaged memories.

As we've seen in the last few chapters, the quest to heal often leads to the power to enhance. As we learn how to repair damaged brains, we'll discover an immense amount about how the brain works. That, in turn, will lead to devices that can improve our mental abilities.

Using electrodes and brain chips, we've already learned a lot about how the brain encodes three of our senses — sight, hearing, and touch. We've learned which neurons in our brain fire in response to which kinds of images, sounds, and sensations. In some sense, we've begun to crack the code of these functions of the brain. We've also learned the basics of how the brain encodes the intent to move our limbs.

The next frontier is to discover how the brain encodes our higher functions — language, memory, attention, emotion, and so on. We now have the technology to study the brain and discover exactly these things. If we can learn these neural codings, if we can crack the code of our highest mental functions, then we can manipulate them. We can repair them when damaged. We can, potentially, improve on them - giving ourselves better memories or new mental abilities. We may be able to control them - giving ourselves the ability to alter our feelings, thoughts, and perceptions. We may even be able to transmit some of the private contents of our brains - our thoughts, memories and emotions - from one person to another.

So far that's science fiction. We have just begun to take our first forays into repairing damaged brains and transmitting information in and out of them. It will take years of painstaking research to decipher the neural codes of the higher brain functions. Along the way there will be mistakes, human trials that harm rather than help their patients. Even once neural prosthetics are routinely employed to repair damaged brains, it will require a revolution in surgery before they're safe enough and convenient enough to entice perfectly healthy men and women to receive these implants.

Neural interfaces are worth discussing not because they're right around the corner - they're not - but because they may effect such tremendous changes in our lives. The ability to tinker directly with the inner workings of our brains offers us more power over ourselves than any technology we've discussed thus far. It offers us the potential to control our emotions in realtime, to rescript our personalities, to communicate our innermost thoughts and feelings to one another, even to draw on the native abilities of our computers as if they were our own. These abilities will pose serious questions to our senses of identity and individuality. They will blur the line between man and machine. They may even blur the line between one person and another. More than genetic engineering of our children, more than the ability to genetically and pharmacologically alter our minds and bodies, more even than the conquest of aging, the integration of our minds with our computers and - through them - each other will challenge our ideas of what it means to be human.

For now, medical research into using brain implants to assist the blind, the paralyzed, and those who've suffered brain damage will drive the development of the technology. It will propel the design of more sophisticated implants and safer surgical procedures to implant them. It will teach us more about how the brain works, and it will gradually introduce the idea of brain implants to the general public, reducing the shock that will come when people begin using these implants to enhance themselves rather than cure disease.

With this path in mind, it's time to look more closely at the history of brain-computer interface research and what it tells us about the future.

Robotic Limbs

Research on restoring motion to the paralyzed hasn't stopped with Johnny Ray. Phil Kennedy and his collaborators have implanted electrodes into the motor cortex of another dozen patients. Kennedy's latest studies are using eight electrodes rather than just the one that Johnny Ray received, allowing their patients to learn to use the onscreen typing system more quickly and easily. Animal studies have gone even farther. And research in the area has been boosted by funding from an unexpected source - the military.

DARPA, the Defense Advanced Research Project Agency, is the US military's premier science agency. It's chartered with conducting and funding basic research that will give the US military a cutting edge in battle, and which will also find its way into civilian use. Odds are you've already benefited from a DARPA research project - the Internet is an outgrowth of something called ARPANET, which DARPA's predecessor created as an experiment in computer networks that could survive a nuclear war.

In 2002 DARPA gave $24 million to a team of researchers led by Miguel Nicolelis of Duke University and John Chapin of the State University of New York. Nicolelis and Chapin are working on brain computer interfaces to give the paralyzed control of robot arms. DARPA's interest is in giving US soldiers the ability to control tanks, fly planes, and share information purely through thought. By cutting the hand out of the system, DARPA hopes that they can improve reaction time and pilots the ability to control more complex systems than their hands can currently manage.

Miguel Nicolelis came from Brazil to Hahnemann University in Pittsburgh in 1989 at the age of 28, with both an M.D. and a PhD from the University of Sao Paulo under his belt. He's a serious looking man, with a thick bushy beard, receding hair, and sad eyes behind large square glasses. He'd always been interested in decoding the function of the brain. At Hahnemann he met

John Chapin, already 10 years into his own career and working on ways to record the activity of neurons.

Over the 15 years of their collaboration, Nicolelis and Chapin have produced some of the most impressive results in the field of direct brain control of computers and robots. Compared to Kennedy's work, they've managed to increase the speed and accuracy of using a computer prosthetic, while dramatically reducing the amount of training time required.

Their first breakthrough was in 1999, when they authored a pivotal study that showed that a rat could control a robot arm in much the same way that Johnny Ray could control a computer cursor. In 2000, they showed that an owl monkey could do the same, and added a new twist. Instead of sending the signals from the monkey's brain straight to the robot arm, they transmitted the information about the monkey's neural activity across the internet to a lab at MIT, 600 miles away, suggesting that humans with neural prosthesis could control robots in remote hazardous environments, or pilot unmanned planes over a battlefield from the comfort and safety of a base.

In their 2003 study, funded by DARPA, the researchers implanted 700 electrodes in the motor cortex of rhesus monkeys - close relatives of ours. They then trained the monkey to control a robot arm using a joystick. As the monkey controlled the arm, a computer recorded the activity of its neurons.

Using a special program built for the task, a computer attached to the electrodes found a pattern in how the neural firing related to the movement of the robot arm. The researchers had cracked the neural encoding of motor control. Armed with this code, they could now predict how the robot arm would move based on the activity in the monkey's motor cortex. In trial after trial, their predictions matched what really happened.

Then came the moment of truth. Nicolelis and Chapin left the joystick in place but disconnected it from the robot arm. They gave the computer that monitored the monkey's implant control over the robot arm instead. Now the motion of the arm would depend entirely on the signals picked up by those 700 electrodes, and not the way the monkey moved the joystick. Would the code work? It did. The monkey was now moving the robot arm purely by thought.

In fact, their processing of the code was a bit too rapid, and the robot arm moved before the monkey expected it. While nerve signals move at a relatively slow 100 meters per second, signals in electronic circuits travel at close to speed of light, about 3,000 times faster. So not only had Chapin and Nicolelis predicted the way the monkey would want to move the robot arm based on its neural activity; they had cut out the delay of sending those signals from the monkey's brain to the muscles that controlled its hand.

Something even more surprising happened next: in some trials, *the monkey stopped using the joystick at all*. It let its own hand drop to its side, and relied solely on the neural interface to control the robot arm. According to Nicolelis, "the monkey suddenly realized that she didn't need to move her arm at all. Her arm muscles went completely quiet, she kept the arm at her side and she controlled the robot arm using only her brain and visual feedback. Such findings tell us that the brain is so amazingly adaptable that it can incorporate an external device ... as a natural extension of the body. "

Such a natural extension has interesting consequences. Before the implant, the monkey had four limbs - its two arms and two legs. Now it had control of *five* limbs - two legs, two biological arms, and one robot arm. This is the kind of thing that gets DARPA excited. It suggests that a pilot with an advanced brain implant might be able to do more things at once than a pilot using just his two arms and two legs. Human attention, rather than our number of limbs, may very well be the bottleneck. But the very hint of an advantage has DARPA interested in follow up studies.

Nicolelis and Chapin's work represents three other advances over the implant the Johnny Ray received. First, tapping into the activity of more neurons increases the amount of information coming out of the brain. It increases the system's bandwidth - the speed and strength of the connection between brain and computer. By using hundreds of electrodes instead of just one, Nicolelis and Chapin were able to give the monkey much quicker and more accurate use of the robot arm than Johnny Ray achieved with his cursor. Second, the new prosthesis required very little training. The monkey learned to use it in about two days, as opposed to the months of training that Ray went through. This is partially a result of the higher bandwidth between the monkey's brain and the computer. It's also a result of a computer program that did some of the learning itself, rather than relying entirely on the brain to adapt. Finally, in previous experiments, Nicolelis and Chapin showed that they could also send touch information back into an animal's brain. That's a necessary step for a useful prosthetic arm in humans.

In 2004, the researchers tested a version of their system on humans undergoing brain surgery to implant a device to control Parkinson's disease. Nicolelis and Chapin had a window of just five minutes with each Parkinson's patient to test their techniques. In that time they implanted a temporary device to monitor the patients' neural activity, and then showed each patient how to play a simple video game using a computer joystick. While the patients played the game, the device's array of 32 electrodes analyzed their patterns of brain activity and software tried to correlate their neural signals with the motions their hands were making on the joystick. Amazingly, after just a few minutes each, the system was able to find a pattern, offering the best evidence yet that a high bandwidth neural interface could immediately assist paralyzed victims.

Nicolelis's team has received FDA approval to run more human experiments. At the same time their success has sparked interest and competition in the field. More than a dozen universities and at least 3 private companies are now working in the area. Many of them have produced results almost as impressive as the ones we've just discussed, and several of the teams plan human trials in the next few years.

Restoring The Senses

Before Phil Kennedy implanted an electrode in Johnny Ray, William House was working on restoring lost hearing. In the 1960s, House implanted the first cochlear implants in profoundly deaf patients, restoring them to partial hearing.

There are 34 million deaf or hearing impaired people in the United States. Normal hearing aids work by amplifying sound that they pick up and then relaying it acoustically to the inner ear. In other words, they turn up the volume on the outside world. For most of those 20 million people, this sort of hearing aid can help. However, for the 2 million Americans who've lost the hair cells in the inner ear that stimulate the auditory nerve, turning up the volume does not help. In these people, there is nothing left in the ear to pick up the sound vibrations and turn them into neural impulses.

Today, more than 70,000 patients worldwide have cochlear implants. A modern cochlear implant typically has 22 electrodes. The different electrodes respond to different frequencies of sound. The more sound an implant picks up in one of those ranges, the more rapidly the corresponding electrode fires. For example, when an implant microphone picks up extremely low frequency noises, it triggers the first electrode. The higher the volume of the low frequency noise, the more rapidly the low frequency electrode fires.

The auditory nerve that cochlear implants stimulate, by contrast, has more than 30,000 fibers, each of which can carry a separate signal. A single electrode can stimulate somewhere between one and ten such nerve fibers. Knowing this, no one would expect that a 22-electrode implant would be able to produce useful hearing, but they do. Cochlear implant patients can generally pick up about 90% of two syllable words by hearing them alone, and more from context and by watching the lips of the speaker. This isn't perfect hearing, or even very good hearing, but it's a marked step up from deafness. It's also surprisingly good considering the tiny number of electrodes compared to nerve fibers.

We'll see this again in every other kind of neural interface we look at: just one or a few electrodes is enough to produce useful results, even when the brain area to which the electrodes connect contains tens of thousands or even tens of *millions* of neurons. In part, we have to credit the amazing adaptability of the brain for this. The human brain seems remarkably good at interpreting and pulling information out of any regular signal. As we'll see, this quality of the brain is a major benefit to brain-computer interface research.

For those few patients who have suffered hearing loss due to damage to the auditory nerve itself, cochlear implants do not solve the problem. For those patients, however, there is ongoing research into implants that would send signals directly into the auditory cortex rather than through the auditory nerve. Initial results seem promising, though the technology is far from being commercialized.

Cochlear implants demonstrated for the first time that electronic devices can be usefully integrated with the brain, even with only a single electrode. Building on this success, other researchers have gone on to work on the other senses - especially vision.

In September of 2002, CNN news showed a video clip of a man named Jens, a fit looking 39-year old Canadian who chops wood and plays the piano, driving a Mustang convertible around in an office parking lot. The only unusual thing about the video was Jens himself - he's blind, and had been for nearly 20 years. That was before he met Dr. William Dobelle.

There are 1.3 million blind people in the US, and millions more worldwide. Dobelle has been working for 30 years to build a system that can restore sight to at least some of them. He does this by implanting electrodes in the visual cortex of the brain and connecting them to a computer. The computer, in turn, is decoding video signals from a camera mounted on the patients glasses, and turning them into electrical impulses to send down the electrodes and into the neurons of the brain. Dobelle's patients don't get perfect vision, they don't even get very good vision today, but what they get is enough to be useful - it was enough for Jens to drive that Mustang around without hitting the lamp poles in the parking lot.

Dobelle's announcement and the subsequent press attention caused an uproar in the field — not because of his research, but because of how he went about it. Unlike the vast majority of brain-computer interface researchers, Dobelle is not in academia. He seldom publishes results, preferring to keep his methods secret in the hope of patenting them. He circumvents the normal medical and ethical review boards and FDA approval by having the surgery to implant his devices done overseas. And unlike other trials of experimental techniques, which are typically free to patients and supported by government grants, Dobelle charges patients for the procedure, which he says costs around $100,000.

Still, it's hard to argue with Dobelle's results. He's given us a remarkable proof of principle. What we've learned is the profound and important result that *it's possible to stimulate the brain to produce useful vision*. If other researchers can improve on Dobelle's work and bring the costs down, such techniques might someday be used to cure blindness in millions of patients.

The focus of visual prosthetic research today is restoring normal vision. But everything about the technology lends itself to someday providing *super*-normal vision. The implant Dobelle inserted in Jens' brain gets its signal from an external video source. Right now that video comes from a camera worn by the patient. In principle, though, the video can come from any source to which the patient has access. To the computer and the electrodes in Jens' brain, it's just another video signal, regardless of where it comes from — a computer, a DVD player, a video game console, a virtual reality system, a remote surveillance camera, or maybe even a camera worn by someone else. Those video devices, in turn, could have abilities beyond those of the human eye. Want digital zoom? Dobelle built this feature into his first visual prosthesis, for a patient named Jerry. Want infra-red or x-ray? Want a visual representation of sonar? All of these are possible for a human with a digital video input interface.

The most important thing Dobelle has done is start to crack the code of vision, in much the same way that Kennedy, Chapin, and Nicolelis have made progress in cracking the neural code of motion. Prior to Dobelle's work, scientists had a general idea of how the brain encodes sight. Now we have confirmation that we know the code well enough to put a crude image into someone's mind. That code is now a kind of video format. Today we already have different video formats for television, digital television, DVDs, and several different types of computer video. Now "neural video format" is added to that list. Given an image you'd like someone to see, Dobelle can figure out what pattern of electrical impulses to send into the visual cortex. Given a whole series of images - say, the frames of a movie - he can do the same. And just as we can convert from any existing format to any other existing format, we'll be able to convert any video that exists today into this neural video format.

Dobelle isn't the only researcher working on visual neural interfaces. There are at least half a dozen groups around the world pursuing this goal. The field is split between two general techniques: connecting directly to the brain versus connecting to the retina. Dobelle is in the direct-to-the-brain camp. His implant sits on the surface of the brain area called the 'primary visual cortex.' The primary visual cortex is the first of several layers of visual processing in your brain. It sits in the very back of your brain, conveniently near the skull and accessible to surgery. Also convenient is that it encodes visual information in a pretty simple manner: patterns of neural activity in this area tend to be shaped like the object you're looking at. If you focus on a large red triangle, a triangular area of your primary visual cortex becomes more active. This kind of simple coding lends itself well to interfacing with an external device.

In the other camp are several research groups working on *retinal prosthesis*. These interfaces are similar in some ways to cochlear implants. Technically, retinal prosthetics don't connect to the brain at all. Instead, they're wired to the retina — the layer of cells that pick up light at the back of the eye — or to the optic nerve that carries information from the eye to the brain. Retinal prosthetics and visual cortex implants each have their respective strengths. The retinal prosthetics are easier and safer to implant, but only work if the patient has a working optic nerve, which patients with severe damage to the eyes may not have. Visual cortex implants work in a wider variety of cases, but at the cost of a more invasive surgery to implant them. Because of their respective advantages and disadvantages, researchers will likely continue to pursue both paths.

Brain to Brain Communication

Getting images and sounds into the brain is one thing, but DARPA has a larger goal. They'd like to see humans able to *send* sounds, images, and other kinds of information to each through direct brain-to-brain communication. As ambitious as this goal is, the basic workings of the brain suggest it's possible.

In 1999, Harvard neuroscientist Garrett Stanley performed an experiment that hinted at this ability. Stanley and his team implanted electrodes in the retina of an anesthetized cat. The electrodes recorded the activity of neurons in the retina of the cat's eye as its head was turned to look at various objects. The information from the electrodes was fed into a computer. Using software Stanley's team wrote, the computer converted the neural signals into a standard video format and sent them to a screen. On the screen you can see what the cat is looking at. The images are blurry, but clearly recognizable. The cat's head is turned to look at the branch of a tree. On the screen, you can see the branch, even if you can't make out all the details of its bark and texture.

Stanley's research is evocative. It suggests, for instance, that we could record experiences at the neural level and play them back later. Just as we could take a recorded video and turn it into a set of nerve impulses for your brain, we could record the actual nerve impulses that are going on, and then electrically stimulate your nerves to cause the same pattern again at a later time.

In principle we could do this for all the senses - record not just what you see, but also what you hear, taste, smell, and feel, all at the level of your brain. Playing back such an experience would be a little like reliving it.

You might even be able to play that kind of sensory recording back for someone else, turning the experience you had into a set of nerve impulses that could be sent into their brain, allowing them to experience at least the sensory parts of an event from your perspective.

Stanley's research is limited though, in that pulls information only out of the peripheral nervous system, not the brain itself. The Harvard researcher showed that he could detect what a cat was seeing at the time, but not what went on deeper in its mind.

A research team at Vanderbilt University plans to go one step farther. Under a $1 million grant from DARPA, psychology professor Jon Kaas and his colleagues are studying the possibility of communicating sounds directly from brain to brain.

In their current experiment, they plan to insert around 30 electrodes into the auditory cortex of two rhesus monkeys. When neurons in the brain of the first monkey fire, a computer will send a signal to electrodes in the matching part of the *second* monkey's brain. The researchers will play sounds for the first monkey. Will the second monkey react to the sounds? Better yet, will the second monkey be able to tell what the sounds played for the first monkey are? The odds are very good that the researchers will be able to communicate *something* between the two. Whether the second monkey will hear an intelligible sound or something completely garbled is a harder question. Yet success with the cochlear implant suggests that it's certainly possible.

Kaas and Stanley's research opens up other possibilities. For example, it's a general principle of the brain that, from a neural perspective, *sensation* equals *recollection* equals *imagination*. What that means is that, for the most part, the same neurons fire if you see something, if you remember seeing it, or if you imagine it. The same is roughly true for sounds, smells, tastes, and physical sensations.

This has interesting implications. It means two humans with brain computer interfaces could send imagined sights, sounds, and sensations back and forth between each other. It also suggests that our imaginations might be useful as input devices for computers. For example, you might imagine a picture and send it to your computer, capturing it more quickly and accurately than you could by drawing it out with your hand. The computer might not even have a video screen. It might simply project the image back into your visual cortex, giving you instant feedback and allowing you to sharpen and modify the picture you're working on inside your head.

Our understanding of the brain suggests this kind of feedback loop between man and machine is possible. We've already seen that we can project images, sounds, and sensations into a human brain, and both basic neuroscience and limited experiments suggest we can send them out. If humans could start collaborating with computers on the contents of their imaginations, it would revolutionize computer interfaces. Instead of having to point and click or type to instruct the computer to do something, we could simply think about it. The ability to communicate with one another in sights, sounds, and sensations, in addition to our current tool of language would have just as profound an impact on human interaction. We'll explore these implications in a few pages.

Deep Brain Stimulation

There are 1.5 million victims of Parkinson's disease in the US. Parkinson's disease sufferers experience uncontrollable trembling of their limbs. The trembling starts as a minor annoyance, and then grows in severity, until eventually the patient is unable to control his or her limbs. There is, as of yet, no cure for Parkinson's. Drugs like L-dopa can control the symptoms of the disease for a while, but in every patient the drugs lose effectiveness over time. At best they delay the progression of the disease.

In some cases, Parkinson's tremors become so bad - and the drugs so ineffective - that doctors cut out part of the brain to stop them. In 1977, a French neurosurgeon named Alim-Louis Benabid was performing such an operation to remove the tremble-producing part of a patient's brain. At the outset he electrically stimulated various parts of the patient's brain to help map its various functions. When he accidentally stimulated the thalamus, the Parkinson's tremors stopped. Since then doctors have been experimenting with the technique with better and better results.

In 1997, the FDA approved deep brain stimulators (DBS) based on Benabid's research for the control of Parkinson's symptoms. A deep brain stimulator is basically a single long electrode that penetrates a few inches through the patient's brain and stops at the thalamus or a nearby area. In

70% of patients with severe Parkinson's, no longer controllable by drugs, a DBS can stop or markedly reduce symptoms. The therapy has proven so effective that it's been implanted in more than 30,000 patients worldwide. In 2002, for the first time, Medicare started paying the $30 - $40,000 bill for the procedure. With the cost barrier to many patients eliminated, doctors expect the therapy to rapidly grow in popularity.

Interestingly enough, in the past few years doctors have noticed that deep brain stimulators can affect more than Parkinson's disease and other movement disorders. In the late 90s researchers noticed that some patients who received deep brain stimulators would experience short periods of depression afterwards. Then they noticed the opposite: some patients implanted with DBS had suddenly improved mood. Even some patients with lifelong depression, something they'd lived with long before Parkinson's disease, were becoming happier and more energetic. There were also reports of patients with severe obsessive compulsive disorder (OCD) who, after receiving a deep brain stimulator to control Parkinson's disease, also stopped showing the signs of the OCD.

Later investigation has shown that it's exactly where the electrodes are placed that determines whether DBS will have side effects, will depress a patient, will lift mood, or will combat obsessive compulsive disorder. Building on this research, at least three groups in the US and several in Europe are now running human trials of deep brain stimulation for severe untreatable depression and OCD. If those trials prove successful, we'll see the first widespread use of brain implants to control depression and OCD around the end of the decade.

Other researchers have seen other evidence that electrical stimulation can affect emotions. In 2003 Takeshi Satow of the University of Kyoto reported that stimulating the inferior temporal gyrus of a patient undergoing surgery for epilepsy induced feelings of happiness. After 5 seconds of stimulation, Satow's patient smiled. With a stronger current, she burst out laughing, for no particular reason at all.

These results aren't new. In the 1960's a pair of scientists working independently of each other demonstrated similar and even more powerful effects on human emotion and behavior through electrical stimulation. Their names were Robert Heath and Jose Delgado.

There's a famous series of pictures of Delgado taken in 1964 Madrid that speaks volumes about his work. In the first photo, Delgado is in a bullring, looking young, dashing, and trim with a full head of black hair. Across the grass from him, still some distance away, is a large bull. In Delgado's right hand is a cape. In his left is a black box with a long antenna. In the second photo, the bull is charging. It's leaning forward, head down, hooves barely touching the ground, horns pointed straight at Delgado's midsection.

In the last photo, Delgado has dropped the cape and pushed a button on the black box in his hand. The bull is skidding to halt, just feet in front of him. Its legs are splayed forward to arrest its charge. Dirt is flying up from its hooves.

Delgado, a Yale researcher at the time, had demonstrated a previously unheard of ability to electrically affect the brain. When he pressed the button on his transmitter it sent a signal to a receiver worn on the bull's head. That receiver, in turn, sent an electrical pulse down electrodes implanted in the bull's hypothalamus, deep within its brain. That pulse shut down the bull's aggression and instilled a sense of tranquility. So long as Delgado continued to send the signal, the bull remained calm. Even efforts by the toreadors to enrage it failed. Conversely, by pressing a different button, Delgado could stimulate a different part of the bull's limbic system, driving it instantly into a rage, regardless of what it was doing at the time.

Delgado did similar work on chimpanzees and other primates. He found that he could electrically control sleep, appetite, sexual arousal, aggression, and even social behavior. In most cases he could change behavior at the flick of a switch. An awake and alert monkey would be asleep within seconds of a current being sent to the correct part of its brain. Its other behaviors could be altered as quickly.

In later studies, Delgado and Tulane researcher Robert Heath independently showed that they could cause similar effects in human psychiatric patients. In a series of ground-breaking and sometimes bizarre experiments, Heath placed electrodes in the brains of some 26 patients suffering from severe mental illness. Those illnesses included incurable epilepsy, Parkinson's disease, schizophrenia, and a variety of psychiatric conditions for which the patients had been hospitalized. Heath's goal was to understand the pleasure and so-called "aversive" systems in the brain. At first he merely recorded activity. He found that when a patient flew into a violent rage, there was a large uptick in activity in the aversive system.

Heath found that he could control those symptoms by stimulating the patient's pleasure areas in the septal region of the brain. This area, part of the limbic system, has since been found to be important in a wide variety of human emotions. By stimulating this area, Heath could calm a patient in a rage, reduce anxiety, and induce smiles and giggles. Many of his patients would suddenly talk about how much they loved their doctor and how wonderful it was to be in the hospital. Heath found that he could also remove severe pain, depression, and delusions. Delgado produced similar results.

Even without Heath and Delgado's past work, there are other good biological reasons to believe that we can affect mood electrically. For example, most anti-depressants work by increasing the amount of serotonin available in your brain. Serotonin is a neurotransmitter used by one neuron to send a signal to the next. In the brain it produces a sense of calm and well-being. Drugs like Prozac and Zoloft prevent neurons from sucking up serotonin from the area around them. With more serotonin bouncing around, there's a better chance that some of it will bump into a receptor on another neuron. The result is a boost in mood. Another way to achieve this might be to stimulate serotonin-producing neurons to release their serotonin out into the brain. The brain's serotonin neurons are concentrated in a place called the Raphe Nucleus. From there they send out long projections into dozens of other parts of the brain. Just one electrode attached to the Raphe Nucleus could trigger those serotonin neurons to release their serotonin. This would have a strong euphoric effect.

Brain computer interfaces may turn out to be far more powerful and precise tools for controlling mood than drugs or gene therapy. Mind-altering drugs are rather clumsy things. Whether swallowed, injected, or smoked, drugs work their way through the blood stream and into the brain. In the brain they spread indiscriminately, flooding every part of complex nerve network. Drugs generally have affinity for only one or a few types of neurons, but they act on *every neuron* of that type.

Drugs also take time to permeate our brains, time to work their way out of our system, and time to cause their effects. Prozac takes several days to elevate mood. A pain killer like Vicodin operates more quickly, but still takes many minutes to an hour before it begins to have its effect. Once the effect has begun, there's no easy way to stop it. The effects will last for several hours, regardless of the desired duration.

Future neural implants, on the other hand, could stimulate one neuron while leaving the neuron next to it alone. They could alter our brains in much more precise, sophisticated ways than the simple floods of chemicals that we use today. Because they're more precise, they might be able produce their effects with fewer side effects. They would also have their effects immediately. They might be used to reduce pain, boost mood, or instill a sense of calm with virtually no delay between activation and effect. When the mood alteration of pain suppression was no longer desired, the person wearing the neural implant could turn it off immediately, stopping the flow of current to the electrodes and ending the effect, instead of waiting for hours for a drug to be cleared from the body.

Perhaps the most intriguing possibility of neural prosthesis for emotion is the potential for communicating emotions directly from brain to brain. Just as we described recording the neural impulses that happen in your brain when you see or hear something, researchers could monitor the neural impulses that occur in the emotional centers of your brain, and then stimulate a similar pattern (perhaps at a lower intensity) in someone else's brain. Just as we could record the activity in your sensory or motor areas for playback later, the activity in your emotional areas could also be recorded and replayed in the future for you or someone else.

In addition, neuroscientists already know a reasonable amount about the brain areas involved in *empathy*. The amygdala and other parts of the brain are responsible for modeling the emotions of other people, giving you a sense for what they're feeling. Information from the rest of your brain is sent into the amygdale to feed into this model. With neural prosthetics, information from the emotional centers of someone else - say, a loved one - could be piped straight you're your empathy center. So rather than having to guess what your spouse or child is feeling, you would simply be sensing it via the wireless link between your brains. If you wanted to sense other people's feelings less, you could choose to turn down the volume, reducing the strength of the signal that your neural interface sent into your empathy centers. Software could even decide which people's feelings to send into your empathy centers and which not to. The end result might be just like having an unusually keen sense of how others are feeling, with the option to dial that sense up or down in intensity based on whatever critera you chose..

Repairing Memory, Perception, and Cognition

In the 2001 film *Memento*, actor Guy Pierce plays Leonard, a man suffering from a kind of brain damage. Leonard is able to think and reason about the world around him. He's able to keep track of things as they're happening. But he's unable to form new long term memories. A few minutes after something happens it's simply gone from his mind. Leonard's is an extreme example, but such cases - and others more bizarre - do happen.

Every year around eighty thousand Americans suffer a blow to the head that results in permanent damage of some kind. More than seven hundred thousand suffer strokes, not all of which can be fully recovered from. The CDC estimates that in the US alone, more than 5 million people are living with some sort of permanent disability resulting from injury to the brain. These people have problems than span the whole gamut of the brain's function - problems with concentration, attention, memory, and perception; lost ability to produce or comprehend language; reduced ability to plan actions or solve problems; difficulty in controlling mood and emotion; long term personality changes; depression, fatigue, anxiety, and agitation; and many more.

Repairing these higher functions of the brain is a larger challenge than any that brain implant researchers have tackled to date. To some degree we've already cracked the code of vision, hearing, touch, and motion. Yet in those areas we have the advantage of something concrete to map the brain's pattern of activity to. We can compare the firing of neurons in the visual cortex to the image someone is looking at and find a relationship. We can observe neurons in the auditory cortex and see which ones fire in response to certain frequencies and volumes of sound.

The higher functions of the brain are a bit more abstract. We don't really have a good concrete analog for concentration, or perception, or memory. And while we've made impressive progress in electrically influencing emotion, our tools are still crude at best. Learning to repair damage to the brain's higher functions is going to be a long, slow process. Yet it's also going to teach us a tremendous amount about just how the brain and mind work.

Theodore Berger at the University of Southern California, another DARPA grantee, is already working on brain implants to repair damaged long-term memories. Researchers have known for decades that an area of the brain called the hippocampus is important in taking our current experiences and storing them away for retrieval days, months, or years later. People with damaged hippocampi have trouble forming these memories. In extreme cases they're like Leonard. They live day to day with memories of what happened before their brain damage, but nothing since then. Berger and his team hope to correct this with computer circuits that act like neurons in the hippocampus.

Berger doesn't claim to know how the hippocampus works. Instead of trying to understand it, he and his colleagues have simply mimicked its behavior. Over the past decade they've electrically stimulated parts of a hippocampus millions of times and observed the results. Through this they've built a mathematical model of this part of the brain, and a computer chip that responds to electrical signals in exactly the same way that a slice of hippocampus in a lab dish does.

The next step is to test the chip in living animals. In an upcoming study with Sam Deadwyler of Wake Forest University, they'll interface the memory chip with the damaged hippocampus of a primate. The chip itself will sit outside the monkey's skull, communicating with the primate's brain through a smaller set of electrodes implanted in the hippocampus. The goal of the study is to restore the monkey's ability to learn. If it succeeds, then Berger and his team will start to move towards trials in humans that have suffered brain damage to the hippocampus. They also believe their device may help in restoring some of the function lost in Alzheimer's disease.

While human trials are still years away, Berger and Deadwyler are already talking about the possibility of *augmenting* human memory - making it easier for us to learn new facts or recall things learned long ago. If we can understand how the brain files things away into memory, their reasoning goes, we can improve on the process. We may even be able to file away a copy of the memories on a computer or other outside device, enabling perfect storage of our memories for decades to come. For now, enhancing or capturing human memories is science fiction. We're not yet close to understanding how the brain encodes memory. Yet the fact that serious researchers are contemplating it at all is provocative, and studies such as Berger and Deadwyler's are the key to developing this understanding of how memory works.

Even if the Berger and Deadwyler's animal studies fail to restore memory, which is quite possible, they'll still be the most sophisticated recordings anyone has ever performed of the hippocampus. They'll teach us important lessons about how memory works in the primate brain. That in turn will give researchers a better starting point for the next round of research. And if Berger's work produces any useful results at all, it will encourage other labs to begin to tackle damaged memories, attention, language abilities, and more.

While Berger and Deadwyler are working on memory, James DiCarlo at MIT is working on perception. DiCarlo is interested in a mental disorder called agnosia. In agnosia, patients have a hard time recognizing objects. The most famous case of agnosia is described in Oliver Sacks' book *The Man Who Mistook His Wife For a Hat*. While confusing one's spouse with an article of clothing is uncommon, it can happen in extreme cases.

DiCarlo is looking at how the infotemporal cortex of monkeys recognizes types of objects. In an upcoming experiment, he plans to record neural activity in this part of the brain while monkeys look at pictures of cats and dogs, and computer generated pictures that look a little like a dog and a little like the cat. The monkeys have been trained to press one button when they see a cat, and a different one when they see a dog. The trick will come when DiCarlo tries to *change* the monkey's perceptions. While the monkey is looking at a picture that looks more like a dog, for example, the researchers will send in a neural signal like the one produced when the monkey sees a cat. If the monkey hits the button to indicate that it's seeing a cat rather than a dog, the researchers will know they've succeeded. That will open the way to treatments for agnosia and other disorders of perception. It will also open the possibility of rather bizarre, even psychedelic, manipulations of human perception. It would suggest, for instance, that you could look at a table and be induced to recognize it as a living thing rather than an inanimate object.

There may also be some practical applications of the technology for enhancing human cognition. Computers are better at analyzing some kinds of data and spotting some kinds of patterns than humans. A computer can look at a series of numbers and spot a trend that a human never would. A computer can examine tens of millions of moves in a chess game and find an interesting opening. A computer that could interface with human perception might be able to overlay these insights on top of our own, giving us the benefit of its analysis with the feeling of having noticed something ourselves.

We would in effect be more perceptive. At the moment this is just speculation, yet it's an interesting possibility raised by the current basic research into how the brain turns our raw sensations into our *perceptions*.

Other researchers are considering the possibility of neural prosthesis to repair damage to other parts of the brain. Patients with damage to Broca and Wernicke's area have problems with producing or understanding language. Patients with damage to part of the left frontal cortex can't perform arithmetic or higher mathematics. Autistic children have structural differences from normal children in the amygdala and other parts of the brain. No one is yet performing experiments in any of these areas, but for the first time, researchers are talking seriously about the possibility of repairing this kind of brain damage through neural prosthesis. As the field matures we'll see scientists attempting to tackle these problems. As they do, they'll make new discoveries about how the brain handles language, mathematics, and other higher functions. Those discoveries, in turn, will open up the possibility of interfacing with these functions, improving them, and even sharing them between individuals.

Chapter 10 – World Wide Mind

Someday in the future, you take the plunge and have a computer interface implanted in your brain. You're not the type to take this step lightly or quickly, but it's now been ten years since brain-computer interfaces were approved for the general public. Most of your friends have them. You watch with envy as they interact on levels that you can't. You're tired of hearing them tell you what you're missing out on. Still, you're careful. You wait for others to have the procedure. You pore over the safety statistics and most common side effects. You find the model that's least likely to be obsolete a few years down the road. And you choose a doctor that has lots of experience and a good track record in implanting the devices.

You go to the hospital in the afternoon. The anesthesiologist numbs your skull but leaves you awake. As you lay there on the operating table, the doctor makes a tiny hole in your skull. Through this hole she inserts an incredibly light, flexible mesh of electronic circuits. Using tiny tools inserted through the same hole, she spreads the mesh across the entire surface of your brain. Once the mesh is in place, it sends millions of microscopic filaments down into your brain. Each filament is a thousandth of the width of a human hair — much smaller than even the smallest nerve cells. After this, the doctor implants a tiny transmitter and receiver just inside your skull. Finally, she replaces the small amount of bone that she had cut out and coats the area with growth factors to speed the bone's healing. Once this scar heals, there will be no visible sign of your implant at all.

After the surgery, you wait in the recovery room, feeling no different than before. The doctor visits and tells you that the procedure went well and that the interface seems to be working. You rest, eat, and go to sleep. In the morning, training begins. On the first day, the interface customizes itself to the parts of your brain that handle vision. You're shown a series of pictures on a screen. As you look at them, the implant in your brain monitors the activity of neurons in your visual cortex, noting which ones fire in response to which images. After an hour, your implant has a good working map of your brain. Now, after each image flashes on the screen, you're asked to close your eyes. A picture identical to the one you saw on the screen appears in your mind. Occasionally it's wrong in some way and you push a button to signal the discrepancy. By early afternoon, your implant is able to send you images and moving pictures with almost perfect fidelity.

You nap and wake for your second training session. This session takes the form of a game. You control a space ship on the screen. You must avoid obstacles by moving your ship up and down, right and left, forward and back, but you have no joystick or keyboard with which to do this. The computer does not respond to your voice commands. You're told to *think* about the way you want to move. You find that by concentrating you can sometimes clumsily nudge the ship one way or another. You're frustrated, but the technician calms you down. She jokes about how long it took her to master this game and says that you're a fast learner. You relax a bit and return to the game. Control comes slowly at first, and then faster and faster as you get the hang of it. You could still do better with your hands, but you're getting better. Finally, using the skills you developed in the game, you learn how to control your implant itself. At first you learn only how to activate a mental "screen" on which the implant places information. The screen is inside your mind, projected into your visual cortex by the electrodes of your implant. Once you learn how to activate that mental projection area, you learn how to move a cursor around that screen and click on the mental "buttons" the implant projects in your mind.

Late in the afternoon, you're discharged from the hospital. You now have all the tools necessary to explore your new abilities on your own. Your spouse, who already has an implant, takes you home. The next morning, you start another training exercise. You activate the implant and it guides you through training it to project sound into your mind. You slip headphones over your ears and play a selection of sounds and music given to you by the hospital. Just as with your vision, the implant monitors your brain activity in response to the sounds you hear. By the end of the morning, you can use the implant to play sounds in your mind.

In the afternoon, the implant guides you through a set of exercises, training it for touch. You are instructed to touch every part of your body, systematically. You stand, sit, and lie down. You take a hot bath and a cold shower. Through this, the implant learns the specifics of how the sense of touch affects your brain. By doing so it learns how to stimulate your brain to mimic these effects.

The next day, you start the most intensive phase of training the implant – learning to communicate with language. You relax as the implant uses sight and sound to present you with words, phrases, and sentences. You repeat these to yourself. The implant shows you images, which you name. It instructs you to read paragraphs and pages off of your mental screen. It teaches you to press a mental "send" button at the same that time you mentally speak. By the end of the first day, you have the rudimentary ability to "type" words to the implant by thinking about them. The next day, you practice again. By the end of the third day, you can dictate to the implant as quickly as you can think.

On the following day, you learn to communicate with more than words. The implant shows you how to contact other people with implants. You practice sending words back and forth with your spouse. You're stunned by the ability to communicate silently. Then the implant shows you how to send other senses. You send your spouse what you see and hear and feel. You close your eyes and imagine a flower and a romantic tune and send those as well. Your spouse responds in kind. You begin to explore this incredible way to communicate. You find you can project emotions and even some abstract ideas. In one conversation, you flit back and forth between words, images, sounds, and feelings. That night, you and your spouse make love while opening all of your senses and emotions to each other. The intimacy is beyond anything you have known. It's almost overwhelming, and you have to retreat. But you know that you will try this again, and that your relationship will reach a new level of depth.

The next day, you return to work. Over the coming weeks, you find your work productivity greatly increased. You can work faster, more easily, and more accurately with teammates who have implants than you can with the non-implanted. In meetings with other implanted workers, you routinely work using mental diagrams, images, and silent speech. In mixed company, you take mental control of the screen, projecting text and images onto it and changing them with a facility you never would have reached with a marker and whiteboard.

In the evenings, you continue to explore your implant. You learn how to use it to control sleep, appetite, mood, arousal, and even your heart rate and respiration. You gain the ability to remain calm merely by willing it. You learn how to use the implant to navigate the internet, asking questions and receiving answers just by thinking about it. The implant guides you through increasingly complex exercises. You play memory games with numbers, shapes, colors, names, faces, places, and all sorts of objects. You add, subtract, and multiply. You play strategy games and solve mental puzzles. Through all of these activities, the implant studies your brain's activity, learning to interface with your highest functions.

Little by little, you feel yourself getting sharper, as the implant adds to your mental abilities. You phrase mathematical questions to yourself and the implant supplies the answers instantly. You find yourself able to remember twenty digits at a time, where before you could only remember seven. Your attention wanders less at work and you're able to juggle more things in your mind at once. The answers to questions you pose to the internet come back more quickly and sometimes almost before you ask the question. You routinely trade memories and experiences with other implanted humans. You learn to view the world through other people's eyes. You let others see through yours. As the months and years pass, you increasingly view your implant as a vital and natural part of you. Using it becomes as natural as breathing. You can no longer imagine a disconnected life.

Science Fiction to Reality

This scenario is science fiction, yet everything in it is grounded in the research we discussed in the previous chapter. What we've discovered about the brain so far suggests that in the coming decades we'll be able to link our minds to computers in deep ways, and through those computers to each other.

At the same time, it's important to separate the incredible potential of brain computer interfaces from the here and now. Blindness and paralysis will not be cured in 2006 or 2008 or even 2010. We won't be mentally transmitting our thoughts to each other in 2015. It took twenty years for the cochlear implant to move from the first experimental successes to FDA approval for mainstream use. It took just as long - from 1977 to 1997 - for research into electrically suppressing Parkinson's tremors to make it into a device approved for mass use.

There are two key hurdles that brain computer interfaces need to leap before they'll be widely employed to combat blindness, paralysis, memory loss, and the other problems they've been shown to be effective against, let alone adopted by the healthy masses as a new communication tool. The first hurdle is that of information flow. Current systems in use in human beings can interact with anywhere from one neuron, in Johnny Ray's case, to 256 neurons, in the case of the visual cortex prosthesis that's partially restored Jens' vision. It's amazing that these systems have produced useful results at all, considering that both vision and motor control involve *billions* of neurons.

To go from proof of concept to real usage, though, future implants will need to connect to many more neurons. Jens sees in a 16 pixel by 16 pixel grid. By way of comparison, a normal television broadcast is a grid 480 by 440, for over 200,000 separate pixels. A good computer screen has more than a million pixels. To provide Jens with something closer to normal human vision, researchers will need to connect to more neurons at a time.

Fortunately, that's quite doable. Neural interfaces have such low bandwidth because they're built on decades-old technology. For most of the field's history, it's been populated by neurosurgeons and neurobiologists - scientists who understand the brain, but aren't on the cutting edge of electronics. Over the last few years, researchers with backgrounds in computer science and electrical engineering have entered the field of brain-computer interfaces. They've brought with them the latest expertise in fabricating circuits and a goal of greatly increasing the number of connections implants make with the brain.

Researcher Ken Wise at the University of Michigan, for example, is producing implants fabricated like microchips. Wise's lab is just finishing their 1024-channel wireless implant. It measures just 3 millimeters on a side and communicates wirelessly through the skull at 1.5 megabits per second. German firm Infineon is working on a neuro-chip that can connect to 16,384 neurons at a time.

These devices are a major improvement from current systems, but they still have plenty of room to improve. Intel's Pentium 4 processor is not much larger than existing neuro-chips and has 55 *million* transistors on it. It has about forty times as many circuits as the optic nerve has neurons, in about the same cross sectional area. Circuits built by the latest chip fabrication techniques are actually much smaller *and* more tightly packed than neurons in the brain. And most of the basic technology used to build Pentium chips can be applied to building better, higher bandwidth brain implants. As the field of neural prosthetics matures, it will borrow more and more of these techniques from the computer industry, enabling the creation of brain implants that can connect to millions of neurons at a time. Eventually the limitation will not be the hardware, but the number of neurons that we can physically connect to without damaging the brain.

The more significant hurdle is brain surgery. Cutting open the skull and placing something inside the brain comes sizeable of risks. One out of every fifty patients to receive a deep brain stimulator, for example, suffers some kind of bleeding in the brain afterwards. In rare cases, that can prove fatal. Even putting aside the risk, brain surgery is an expensive process that often takes days to recover from. Deep brain stimulators cost an average $60,000 dollars to install. Cochlear implants cost only a small amount less. Many brain surgeries, such as aneurysm repair, require three or more days of recovery in a hospital before the patient can go home. The combination of these issues will give anyone considering brain surgery pause. If neural interfaces are ever to become mainstream, whether for their medical use or as enhancements, scientists and physicians will have to dramatically lower the health risk, cost, and inconvenience of tapping into the brain.

Researchers are looking at two ways to get around these problems. The first possibility is entirely *non-invasive* interfaces. Instead of actually attaching electrodes to neurons, future brain computer interfaces might be able to monitor and interact with individual neurons from *outside* the skull, without a need for surgery at all. DARPA in particular is pouring money into this avenue of research.

Today there are two top contenders for a non-invasive interface. The first is electro encephalography or EEG. An EEG device works by placing anywhere from one to more than one hundred electrodes on the scalp, held in place by sticky pads. EEG devices pick up the aggregate electrical activity of the hundred billion neurons in the brain. EEG scientists talk about brain activity in terms of "brain waves." Neurons, it turns out, tend to fire with a certain rhythm. Exactly how fast they fire depends on what kind of neurons they are, where in your brain they are, and what you're doing with your mind at the time. The more relaxed you are, the slower your neurons are firing. In deepest sleep, most of your brain is active at around one to four firings per second. 1 to 4 Hertz. As you become more active, different neurons in the different parts of your brain that are being put to use start to fire more quickly. EEG devices frequently see 30 Hertz activity in various parts of the brain, and sometimes faster.

Researchers Nils Birbaumer and Johnothan Wolpow have used EEGs as a kind of brain computer interface for paralyzed patients. They've painstakingly trained quadriplegics to change their brainwave activity. The EEG picks up the change and interprets it as a set of commands. Using this technique Birbaumer and Wolpow's patients have controlled powered wheelchairs and have moved cursors around on computer screens, much like Johnny Ray.

EEG devices are small enough to carry around with you. They're also fairly inexpensive to build. If manufactured in volume they might cost as little as a few hundred dollars each.

Unfortunately, Birbaumer and Wolpow's research seems to be pushing EEG about as far as it can go. Beyond about a hundred electrodes, the scalp is completely covered. What's more, because any electrical signal from a neuron has to pass through the skull and scalp (and usually many other neurons) before reaching an electrode, EEGs produce at best very general information about what's going on inside the brain. With EEG it's impossible to get the kind of precise information about specific neurons and patterns of neural firing that you can get from inside the brain. For that reason, most brain computer interface researchers think EEG won't go much farther than it's gone already.

The other non-invasive technology is functional magnetic resonance imaging, or fMRI. fMRI is the most advanced technique for looking at the brain now widely in use. An fMRI machine can detect changes in blood flow in areas as small as one millimeter on a side. When neurons fire, they draw more blood to replenish their supply of nutrients and oxygen. By detecting blood flow, fMRI detects which parts of the brain are most active. fMRI has three main drawbacks. First, even one millimeter on a side isn't very high resolution when dealing with the brain. In a one millimeter cube there may be twenty thousand neurons. Second, fMRI machines generate huge magnetic fields. Those fields can't be used anywhere near metal objects. As a result, fMRI machines need magnets that weigh several tons to generate those fields. Finally, fMRI is expensive, up to several million dollars per device.

Despite the limitations of EEG, and fMRI, we shouldn't rule out the possibility of a non-invasive interface. Researchers are pursuing many different avenues, including picking up the magnetic fields of neurons, shining infra-red lasers through the outermost layers of the brain to pick up changes in opacity caused by neural firing, and even more exotic methods. In early 2003 scientists used an advanced microscope that fires light into its target to take pictures of single neurons. The technology is fairly bulky and can only see a tiny area at a time, but future generations will likely shrink and increase in ability. Researchers have also used something called transcranial magnetic stimulation (TMS) to *change* the activity of neurons the brain. By applying a magnetic field just outside they skull, they can temporarily scramble the electrical signals in a small part of the brain. After the magnetic field is gone, brain activity in the affected area returns to normal. Scientists use this as safe way to probe the workings of the brain, by effectively turning off a small area and seeing how a subject is able to function. TMS has spatial resolution somewhere between EEG and fMRI. It can only affect whole brain areas at a time right now, not individual neurons. It's low precision means that it can only be used to scramble the activity in areas, as opposed to inducing very specific patterns of activity. Today that limits its use as a type of brain computer interface, but researchers are working on ways to improve its precision.

What's more, research in the area of non-invasive interfaces is being fueled by general medical interest. High resolution non-invasive scans of the brain would prove invaluable as diagnostic tools – able to spot tumors and burst blood vessels and watch brain activity in real time. Because of this general interest, hundreds of millions of dollars are going into research projects aimed at improving the resolution of non-invasive scans, as well as non-invasive manipulations like TMS. Interfacing with individual neurons without opening up the skull is a long shot today, but one that's generated quite a bit of interest.

The other possibility is improving the surgical process itself to the point where it's simple enough, inexpensive enough, and safe enough to be performed routinely. There's precedent for surgical techniques advancing to the point where millions of people have them on a purely elective basis. For example, consider plastic surgery. Liposuction and breast augmentation are major surgeries and are performed hundreds of thousands of times each year.

A better analogy may be corrective eye surgery. As of 2002, more than 10 million people worldwide have had a corrective eye surgery. The vast majority of these people already had perfect vision by way of eyeglasses or contact lenses. Corrective eye surgery was, for them, purely a convenience. The only benefit it brings is freedom from having to wear glasses or contacts.

As recently as 1990, the idea of surgically correcting vision was exotic, even a bit bizarre. The most sophisticated corrective eye surgery of the time, radial keratotomy (RK), involved making several incisions in the eye. It carried high risks and a recovery time of up to a week for each eye. That year there were twenty thousand corrective eye surgeries worldwide. What changed corrective eye surgery was the excimer laser. The LASIK procedure, which utilizes the excimer laser, can be done in about fifteen minutes. Recovery can be as quick as a single day, and the risk of complications is far lower than with RK.

Is there an equivalent of the excimer laser and LASIK for brain surgery? It's possible. A leading contender is endovascular surgery. In an endovascular technique, doctors don't make an incision in the skull at all. Instead, they insert a probe through a hole in an artery, usually in the thigh, and guide it through the circulatory system to the brain. Using magnets placed around the head and real-time brain scans, physicians can then steer the device to the exact point in the brain they want to reach.

Endovascular surgery for the brain is still quite new. It was first attempted in the early 1990s and was approved by the FDA for use in the United States in 1995. One popular use is delivering doses of radiation or chemotherapy directly to the site of a tumor, without having to affect the rest of the brain. The most common use, though, is in the treatment of aneurysms.

An aneurysm is a weak area in a blood vessel of the brain. The pressure of the blood under the weak point causes the vessel to balloon out, stretching it further. Eventually it bursts, causing a bleeding in the brain, sometimes severe enough to be categorized as a stroke.

Today physicians estimate that a staggering 18 million people are walking around with unruptured aneurysms. Prior to endovascular neurosurgery, repairing an aneurysm (either before or after it burst) was a several hour procedure that involved drilling through the skull. It typically required the patient to remain in the hospital for three to four days after the surgery. Endovascular embolizations, on the other hand, can repair aneurysms in as little as ninety minutes, with the patient frequently able to leave the hospital on the same day. Follow-ups a year later show that patients who had aneurysms repaired through endovascular procedures recovered more quickly and had fewer complications than patients who went through traditional brain surgery. As a result, the procedure is becoming more and more popular, with a total of about 125,000 endovascular treatments of aneurysms performed between 1995 and 2002.

In 2002, Rudolpho Llinas proposed a way to use the principles of endovascular surgery to tap into millions of neurons at once. Llinas is one of the elder statesmen of brain science. He helped pioneer the study of how neurons function, serves as the editor in chief of the journal *Neuroscience,* and is the chair of the NYU department of the same name. He has a reputation among his colleagues for cool-headedness.

Llinas' idea, which the NSF has provided a grant to explore, is that we use an existing network of passageways that reaches every neuron in the brain — the network of arteries, veins, and capillaries that supply blood to these neurons. In Llinas's proposal for a neurovascular brain-computer interface, a probe is inserted into an artery in the body and guided up to the brain via magnets outside the body, much as it would be in any endovascular procedure. The probe is actually a collection of a million or so much smaller wires, each roughly a tenth of micrometer thick. As the probe is threaded through the brain's vascular networks, the flow of blood and gentle guidance from the magnets separates the wires from each other, allowing them to be carried into a million arterial nooks and crannies. The smallest capillaries in the brain are about ten micrometers wide, or one fifth the width of a human hair. The wires Llinas proposes to use are one hundred times thinner. Researchers have already created wires that thin - and even thinner - using carbon nanotubes.

The advantage of Llinas' proposal is that one wouldn't need to drill through the skull, or even penetrate the blood-brain barrier to insert the interface. Nor would doctors have to push an electrode through layers of neurons to reach the ones further inside the brain – the network of blood vessels extends everywhere. What's more, the system would be completely removable. It could simply be pulled back out in the same way it went in.

Llinas and his team are currently doing proof of concept tests in animals using just a few wires at a time. If the proposal proves viable, receiving a future brain computer interface may be as simple as a ninety minute outpatient procedure, at which point it would become more a more acceptable option in both medical and enhancement uses.

Safe, easy to install, multi-million neuron brain implants would revolutionize the field of neural prosthetics. They would allow us to restore lost vision and hearing to levels as functional as those normally enjoyed by humans. Paralyzed patients would be able to control robotic limbs and computer interfaces with the same ease and precision that the healthy control their natural limbs with, even going so far as to send realistic touch sensations back into the brain.

Higher bandwidth implants would likely give us more subtle, fine grained control over our emotions than the crude experiments of Delgado and Heath. They may allow us to project images, sounds, and sensations as rich as we can imagine to our computers and to each other, even allowing us to communicate our inner *emotions* to each other. Most importantly, these massively connected implants would provide new insights into memory, attention, language, perception, and the other higher functions of the brain. This, in turn, may lead to implants which *enhance* our existing mental functions.

Non-Zero Minds

Neural interfaces bring together the two great information processing systems that exist today – the brain and the computer. The brain's power manifests itself in the flexibility of the human mind. We have language, politics, science, and culture. We excel in learning new skills and adapting to new situations. We're general problem solvers, able to take lessons learned in one domain and apply them to another.

By contrast, our computers are simple, rigid, specialized things. An email program will never learn on its own to handle voicemail, despite the similarity between the two. Deep Blue, the chess program that beat world champion Gary Kasparov, can't tell you what it thinks the crucial move was. It speaks only in chess moves. It can beat the world champion chess player, but it doesn't know how to play checkers. In fact, a three year old human can do countless things this specialized chess champion can't.

At the same time, our computers can do things that our brains cannot. While humans can do complex, general things well, computers specialize in doing simple things at incredible speeds. They can calculate, sort, search, filter, and transform simple, rigid information millions of times faster than humans, without ever becoming bored, tired, or inattentive. They can store and access vast amounts of data and communicate it at phenomenal speeds, without ever understanding what it means.

The capabilities of brains and computers are complimentary - that's why we build computers. The more tightly we can couple those computers to the information processing that happens in our minds, the more we'll benefit. The evidence suggests that we could couple these new abilities and our old ones quite closely. Our brains already have an evolved facility with tools that lends itself to the incorporation of computers as parts of our self image. The last few decades of neuroscience have shown that the brain re-organizes itself in response to reading, to playing the piano, or to other interactions with the tools and devices we've built. In an interview after his most recent primate robot arm study, Miguel Nicolelis said:

> "Our analyses of the brain signals showed that the animal learned to assimilate the robot arm into her brain as if it was her own arm... Such findings tell us that the brain is so amazingly adaptable that it can incorporate an external device into its own 'neuronal space' as a natural

extension of the body. Actually, we see this every day, when we use any tool, from a pencil to a car. As we learn to use that tool, we incorporate the properties of that tool into our brain, which makes us proficient in using it."

Phil Kennedy observed the same thing in Johnny Ray, the first human to control a computer through a neural interface. While Johnny struggled at first with the interface, within a few months he moved it naturally, without even thinking. Six months after the surgery, when Kennedy asked what it felt like to move the computer cursor, Ray spelled out N-O-T-H-I-N-G.

That level of integration suggests that neural prosthesis that affect our higher cognitive abilities will also feel like extensions of ourselves, rather than external devices. If so, we'll be able to marry our strengths of learning, creativity, and especially very *general* thinking and problem solving with the computer strengths of amazing speed, precision, storage capacity, and very *specific* and narrow information processing. We'll be able to take advantage of silicon speed, accuracy and storage in the domains that our minds are uniquely suited to - problem solving, language, invention, art, culture, and emotion. We'll achieve something of the best of both worlds.

Fundamentally, neural interfaces are a new information technology, and the history of information technologies shows that they frequently have their largest impact through their effects on communication. Neural interfaces will give humans the ability to communicate in an integrated stream of language, images, sounds, sensations, and emotions. We'll combine these different modalities just as fluidly as we combine speech and hand gestures today, but to much greater effect. One friend telling another about a great date may tell a story in words, images of her handsome escort for the night, the rich sound of the symphony playing, the taste of the wine they drank, the feel of his hand on hers, and the emotional rush of their goodnight kiss.

Those new communication abilities could scale as well to groups as to small pairs. When sensations, emotions, and ideas become digital, it's as easy to share them with a dozen friends, or a thousand strangers, as it is to send them to one person. Just as we can email our words, or place them on websites, or send them in instant messages or chat rooms, we'll be able to broadcast the inner states of our minds. When and to whom we choose to share such things will be up to us. The technology will give us the flexibility to do what we like with the contents of our thoughts, feelings, and imaginations, and society will respond with new social norms to guide our choices.

Mind-to-mind communication will find applications in the workplace as well. Imagine two engineers planning a project together. As they talk they project mental images to one another - crude schematics of the structure they plan to build. They produce their mental drawings quickly - more quickly than they could draw with pencil and paper or mouse and keyboard. Computer software mediates their communication, capturing the images, allowing both of them to work on editing the plan at once, and storing their work for future reference or for publication to others.

As the engineers mentally sketch, the computer assists their work. It runs calculations on strength and weight, supplying the information as necessary. It rotates the image more fluidly than a human mind can, letting the engineers see the structure from all sides. It stores the image in more detail than the humans can hold in their minds at once. As each engineer focuses on part of the structure, embellishing and clarifying and refining, the computer integrates it with the whole. In effect, it increases the power of their visual imagination, just as would a drafting board, but far more quickly, intimately, and intuitively.

The applications are as important in science. Albert Einstein said that he owed his discovery of relativity to his ability to visualize space and time. Future Einsteins may discuss their work with their colleagues as much in mental imagery as in words. The idea of a "thought experiment" becomes more powerful when thoughts can be projected to others, or assisted by software and hardware outside the mind.

There are similar implications for the arts and media. We have never known "multimedia" of the sort that we'll know with direct connections between our minds. Beyond movies we'll see immersive media that touch all of our senses, our emotions, and more. Direct neural connections to the sensory portions of our brains would allow us to build virtual worlds with one another that are as far beyond the crude virtual reality goggles available today as movies are beyond comic books. Mature neural prosthesis would affect every type of mass communication, from movies to poetry, from music videos to educational texts, from live concerts to museums. All would be enriched by the ability to communicate in images, sounds, sensations, and emotions.

The closest parallel in history may be the event that ushered in the Renaissance - the development of the printing press by Guttenberg in the 1450's. Prior to Guttenberg, books were rare objects, limited to the extremely wealthy. To spread the information stored in a book, a scribe had to manually copy it into another bound volume. Human copiers kept the cost of books high while also introducing errors into each successive copy. The printing press changed that. For the first time, humanity had an efficient way to precisely share large amounts of information. Men and women who had never met, or did not even live in the same years, could share their thoughts with each other in a structured and durable fashion.

In addition, the printing press ushered in a surprising decentralization of society. Prior to Guttenberg, the most powerful entity in Europe was the Catholic Church. The first use of Guttenberg's press was to make copies of the Bible and other religious texts. Yet the press also allowed Martin Luther to publish his pamphlets attacking the Church, thus making the Protestant Reformation possible. The press did help the Church spread its own message, but it had a larger effect in magnifying the voices of those who were not yet being heard.

Over the course of history, information technology - the printing press, computers, phones, and the internet - have generally empowered individuals and undermined dictatorial states. They've facilitated change rather than control. In addition to the Protestant Reformation, fax machines and photo copies helped lead to reform in the former Soviet Union, and internet access is slowly facilitating change in China today (despite attempts at censorship by the state). In contrast to centrally controlled information technologies like television, newer technologies like the world wide web, cell phones, and instant messaging are *distributed*. They allow any person to share information with any other person, rather than simply transmitting the party line. Neural interfaces would be the pinnacle of distributed communication devices, allowing each individual to transmit her thoughts, senses, emotions, and knowledge to any similarly equipped person in the world. History suggests that this development will empower individuals and groups, and undermine central authority.

While undermining central authority, information technology nevertheless increases the benefits of human cooperation. In his book *Non Zero*, Robert Wright proposes a fascinating theory of human history that explains this. Wright draws upon a mathematical field known as game theory. In game theory there are "zero sum" games where one can only benefit at the expense of another. There are also "non-zero sum" games, where the total benefit and harm to the parties involved does not have to balance out. In non-zero sum games, each player can benefit, without any need to harm the others. Looking over 20,000 years of human history, Wright comes to the conclusion that mankind has repeatedly developed new ways to communicate and coordinate our activities to make our interactions more non-zero sum. That is to say, we've invented new ways of working together so that we all benefit. The printing press is a perfect example of this. An author who puts down his work in a book loses very little, and may indeed gain for himself. At the same time, the entire rest of the world may gain tremendously by consuming the knowledge he shares. The total gain for the world far exceeds the loss or effort put in.

The net effect of information technology is an increase in global intelligence. We individuals are, in a sense, like neurons in a global brain - a marketplace of ideas, experience, and innovation. The more power we gain to communicate with one another, the more integrated that brain becomes. In the last few centuries we've taken tremendous steps, from small isolated pockets of computation in individual tribes and civilizations, to the world wide web where anyone with a computer can publish text, images, sounds, and videos to a billion others around the world. The next step is the integration of our biological brains – unlocking the inner ideas and experiences we have, and allowing us to share them with each other, to weave them together into thoughts in a world wide mind.

Chapter 11 – Life Without Limits

Altamira cave sits just inland of the northern cost of Spain, five kilometers from the scenic town of Santillana del Mar. Green fields and gently rolling hills dot the countryside. The historic buildings and cobblestone streets of the town date back to the sixteenth century.

In 1868, a hunter following his dog stumbled across the cave. For the next decade other hunters used it as a refuge from the elements. Yet its importance would be discovered by a young girl. Altamira sits on what was then the estate of Don Marcelino Sanz de Sautuola, a Spanish nobleman and amateur archeologist. In 1879, Marcelino brought his nine year old daughter Maria with him on an exploration of the cave. While he looked down at the floor of the cave searching for bones and arrowheads, it was she who pointed up at the ceiling and exclaimed 'Mira, Papa, bueyes!' (Look, Papa, oxen!). Marcelino and his daughter had discovered the first cave paintings left by Stone Age people.

For almost another twenty years, archeologists would maintain that the paintings at Altamira were an elaborate hoax. Yet by the turn of the century (sadly after Marcelino's death) even the staunchest critics were convinced that the vivid paintings of bison and other animals in the cave had been painted at least ten thousand years ago, well before recorded history.

The discovery of Altamira has been followed by dozens of other finds of cave paintings, burial grounds, jewelry, sculpture, and sophisticated tools dating back a few tens of thousands of years. Archeologists and anthropologists interpret these as the first signs of true culture, the turning point that marks the people who left these artifacts as fully modern humans, possibly even as the first indisputable sign of human language.

This flourishing of culture has a name - The Great Leap Forward - and is dated to around forty thousand years ago. Prior to forty thousand years ago, the archeological record shows primitive stone tools, signs that wooden tools were made as well, and clear evidence for the use of fire. The stone tools are all general purpose cutting instruments. The skeletons look fully anatomically human, with modern sized brains, but there's no art, no multi-part tools, no burial of the dead, or anything else that would separate the humans of this time culturally from those that lived tens of thousands of years earlier.

Suddenly, at about forty thousand years in the past, the archeological record changes. Stone tools become more specialized - instead of simple sharpened rocks there are awls, picks, needles, and other shapes. Bone fishing hooks appear, as do harpoons, spear throwers, and - soon thereafter - bows and arrows. And for the first time cave paintings like those at Altamira can be seen.

The people who left those artifacts forty thousand years ago were identical to us in skeletal structure and probably in DNA as well. Human evolution at a biological level hasn't moved much from that point. But human cultural evolution has moved at an unprecedented rate. Within a few tens of thousands of years of the first cave paintings, our ancestors had domesticated horses and other animals, had discovered agriculture, had built permanent dwellings for themselves where none had been before, and were on their way to discovering metallurgy.

A few short thousands of years after that, they developed writing and mathematics, astronomy and the wheel, philosophy and the beginnings of medicine. Since then, progress has continued to accelerate. In just the last century we've unraveled the mysteries of gravity, of electricity, and light. We've charted much of the visible universe. We've built machines that can manipulate matter and energy to move us from place to place, to shelter us from the elements, to make distant objects seem close, to send our words and images to one another, to store and manipulate huge amounts of information, to perform acts of physical labor that would have been the work of entire nations in the not-so-distant past. Through indirect means, we've already enhanced ourselves.

In these past few centuries we've learned as much about the world inside us as the world around us. We've discovered that our bodies are composed of cells, which themselves are made up of proteins which are described by genes. We've learned that our brains are made up of special sorts of these cells - neurons - and how they communicate with one another. That knowledge has allowed us to heal the sick and is on the verge of allowing us to enhance our minds and bodies.

Every century, every decade, every year things seem to change more quickly. It took tens of thousands of years to go from cave paintings to agriculture. It took thousands of years to go from agriculture to writing. It took thousands more to go from writing to the first steam engine. Yet it took less than seventy years to go from the first automobiles to humans walking on the moon. It was just twenty years between Watson and Crick's discovery of the structure of DNA to Cohen and Boyer's first success at genetic engineering, and less than thirty more to the complete mapping of the human genome.

Now that same accelerating pace of change is on the verge of touching us in the most intimate way - by giving us the power to reshape our own minds and bodies.

There are those who want to put the brakes on this kind of change. Environmentalist Bill McKibben is one of them. If we change our minds and bodies as we've change the world around us, he fears, we'll lose sight of the very meaning of our lives. In his book *Enough: Staying Human in an Engineered Age* he writes, "We are snipping the very last weight holding us to the ground, and when it's gone we will float silently away into the vacuum of meaninglessness ".

McKibben's fear is that we are defined by our limitations. If we exceed them, we will no longer be human. That's why he wants us to say 'enough'. Yet throughout our history we've exceeded our limits and added to our capabilities. If our limits define us, then we stopped being human a long time ago, when we invented tools and language and science that extended the powers of our minds and bodies beyond those our hunter gatherer ancestors were born with.

Leon Kass seems to see things more clearly. In a passage urging respect for our given state, he writes:

> "The human soul yearns for, longs for, aspires to some condition, some state, some goal toward which our earthly activities are directed but which cannot be attained in earthly life. Our soul's reach exceeds our grasp; it seeks more than continuance; it reaches for something beyond us, something that for the most part eludes us."

History suggests that Kass is correct. We have always sought beyond our grasp. This continual yearning for more that he fears is, in fact, a hallmark of humanity. It has driven the evolution of our culture since the Great Leap Forward forty thousand years ago, and continues to drive our progress today.

Only in mythical utopias or dystopias do we ever see a permanent sense of contentment. In the real world, contentment is transitory. It comes with each accomplishment. If it lasted too long, it would stunt our urge to grow and change.

This hunger, this reach that exceeds our grasp, this aspiration to something "which cannot be attained in earthly life," has built our world. It has built our comfortable lives. It has produced the medicine that keeps us alive. It has built our vast store of knowledge. It has produced our art, and music, and philosophy. It has built our deepest understanding of the mysteries of the universe. To never say enough, to always want for more – that is what it means to be human.

We owe everything about our current lives to our predecessors on this world, the ones who would not stand for "enough" and instead asked "what next?" Those brave, foolhardy inventors and explorers sought out better ways to live, ways to live in more comfort, and with greater health, and with greater opportunities for their children. Their curiosity, their willingness to experiment with the unknown, their bravery in the face of risk has brought us everything – from agriculture to written language to the use of fire to antibiotics to telephones and more. The debt we owe them can never be repaid– they're gone from this world, beyond the reach of our gratitude.

If we can never repay that debt, we can perhaps pay it forward to future generations. We can't perfectly predict, may not even be able to imagine, the ways that our descendents will put the technologies we develop to use. Yet the lesson of history is that they will use the powers we provide them to make their world a better place. Now it's time to do what past generations did for us – to explore the world, to experiment with new ways of doing and being, so that we can pass what we learn on to the future - not to try to dictate the lives of future generations, but to let countless families and individuals chart their own course, apply their own best judgments, and use their collective intelligence to propel our species into the future.

Destiny of Species

The chain of events that has led us to this point stretches back into the deep history of our world. Six million years ago, our pre-human ancestors diverged from the evolutionary line that would eventually produce our chimpanzee relatives. It wasn't a sharp divergence – we didn't grow large brains or start composing sonnets overnight. It was more than another three million years until we would start to use the first primitive stone tools, and longer until we would master fire and language. Yet in the history of life on earth, the development from our chimp ancestor until today has happened with extraordinary speed.

In those six million years, we've come to stand upright. We've shed our body hair. We've evolved longer lives. Most impressively, our brains have ballooned in size, and with that expansion has come the capacity for tools, speech, writing, and the trappings of modern society which have so recently appeared.

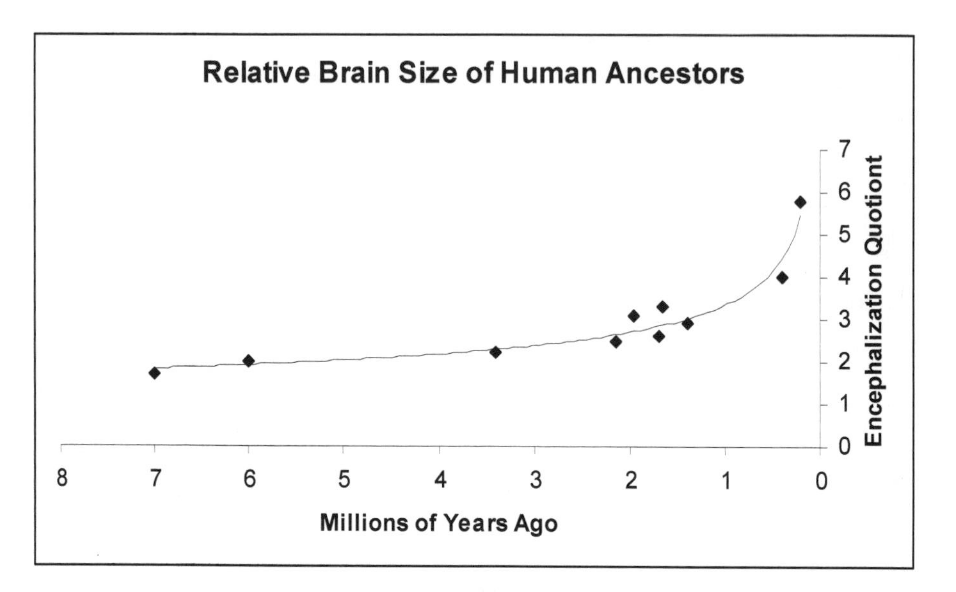

Figure 10 - Human brain size adjusted for body weight has tripled in the last 7 million years, with the trend accelerating over time.

To place this in context, life on earth existed for almost four billion years before the first brain evolved. In the next 60 million years, brains grew to the prodigious 400cc's of our chimp-human ancestor. In the six million years since then, they've more than tripled in size to an average of 1350cc's in a modern human. Something remarkable is going on.

Evolution does not, in general, have a direction. It has no grand plan. It is, in the words of philosopher Daniel Dennet, an algorithm – a simple rule mechanically applied over and over again. If a gene is effective at propagating itself to future generations, it will flourish. If a gene is not effective at reaching future generations, it will die out.

Sometimes and some places, as with recent human history, that process leads to more complex and more intelligent creatures. Other times and places it leads to simplification, and species shed traits they no longer need as evolution selects for the most efficient members. For the most part, evolution leads in neither direction. The dominant form of life on earth, in terms of number of individuals, number of species, and sheer biomass, are single celled bacteria. From an ecological point of view, this is the age of bacteria, as it has been for 3.5 billion years.

While the evolutionary process itself has no direction, the details of a particular environment or species may impose one, at least for a period of time. In a volcanic crater, any life that evolves will be resistant to heat. In Antarctica, evolution will select for resistance to cold. A species that gives birth to only a few children will evolve to invest heavily in each of them, while a species that lays thousands of eggs will evolve to invest less in each.

Some combination of such factors - both environmental and biological - came together in pre-human history to propel the evolution of one splinter off the primate branch of the tree of life towards large brains, sophisticated tools, rich communication abilities, and long lives. There are any number of explanations. Evolutionary psychologists propose that we exist in a "cognitive niche", a particular area in the evolutionary landscape that we've been progressively selected to better fit. University of Washington neuroscientist William Calvin has proposed that frequent and severe climate change over the last three million years selected for creatures with big brains who could easily adapt to new conditions. Biologist George Miller, in his book *The Mating Mind*, theorizes that sexual selection - the choice of who to mate with - rather than natural selection of the survival sort, has driven the development of minds.

We don't know for certain what triggered this evolutionary surge. What we do know is that it marked a phase shift in our evolution. Suddenly six million years ago change in our branch of the tree of life accelerated and new complexity exploded onto the scene. This was not the first such transition in evolutionary history, nor will it be the last.

Life on earth began almost four billion years ago. For the first three billion of those years, all life was single-celled. Only about 700 million years ago did multi-celled creatures evolve. And with their presence on earth came a huge flowering of the varieties of life. The Cambrian Explosion which took place around 570 million years ago produced all the basic body plans of creatures found on earth today in just a few tens of millions of years. For some reason, after almost three billion years in which evolution produced little more complexity than algae or bacteria, everything changed in the blink of an eye. New complexity and new diversity appeared. One branch of life on earth experienced a paradigm change - from single cells and simple colonies, to complex creatures that could walk, and fly, and swim, and even crudely contemplate the world around them.

We humans represent the next such phase shift. Our arrival on this world is as significant in biological terms as that of the first multi-celled creatures. We are as different from every past life form on this planet as chimpanzees are from bacteria. We alone possess the power to alter our own minds and bodies and those of our children. We alone possess the power to guide our own development - to choose our paths, rather than allowing nature to blindly select for the genes that are best at spreading themselves.

We are not the end point of evolution - there is no such thing. We are just an intermediate step on one branch of the tree of life. But from this point on we can choose the directions in which we grow and change. We can choose new states that benefit us and benefit our children, rather than benefiting our genes.

We will not all opt for the same changes. We will not all choose the same direction of travel. Different men and women, different communities, different ideologies will all select different goals to work towards. Some of us will choose to stay as we are, while others will choose to transform. Humanity will expand, splinter, and blossom.

At some point, one hundred years or one thousand years or one million years from now, our world and perhaps this corner of our universe will be populated by descendents that we might not recognize. Yet they will think, and love, and dream of better tomorrows, and strive to achieve them. They will have the traits most dear to us. They will be different in ways we cannot imagine.

Most of all, these descendents of ours will be fantastically diverse. Where we are all alike in the basic forms of our minds and bodies, they will exist in a plethora of forms. Humanity will have given birth, not to one new species, not to a dozen new species, but to thousands, or millions. We will have spawned a new explosion of life as sudden and momentous as the Cambrian Explosion. Long after we are gone, after there no longer live any beings with our DNA, our distant descendents will still look back. They'll look at this moment in our history and marvel. *To think that such primitive creatures as homo sapiens could give birth to whole new kingdoms of life! They could not have understood what they were creating! Yet we're lucky that they had such a strong urge to change and grow – that is why we're here.*

We are, if we choose to be, the seed from which wondrous new kinds of life can grow. We are the prospective parents of new and unimaginable creatures. We are the tiny metazoan from which a new Cambrian can spring. I can think of no more beautiful destiny for any species, no more privileged place in history, than to be the initiators of this new genesis.

Acknowledgements

This book wouldn't exist without the efforts of a large number of people who helped me cross the gulf between the idea and the realization. Especially for a first time author, the encouragement, mentorship, and support I received from friends and strangers alike was invaluable.

At the very beginning, Leo Dirac, and Todd Kopriva gave me much needed encouragement and constructive criticism of my writing. Without them I doubt I would have gotten past the first chapter.

A number of other authors, some of whom I knew and others who I approached from out of the blue, took the time to give me advice on writing a book or to critique what I'd written already. I'm especially grateful to Scotto Moore, Erik Davis, Mark Pesce, Charles Mann, Greg Stock, Robert Wright, and David Brin.

I count myself lucky to have a large circle of friends and associates who were willing to read early drafts or excerpts and provide their critical feedback. Among them are Adam Henderson, Alexis Carlson, Anders Sandberg, Ania Mitros, Antoun Nabhan, Aubrey de Grey, Brent Field, Brett Paatsch, Chris Phoenix, Daniel Westreich, David Lockhart, David Perlman, Elizabeth Kister, Emily Rocke, Gabriel Williams, James Hughes, John Rae-Grant, Lesley Carmichael, Lori Kingery, Mason Bryant, Natalia Powers, Marcelo Calbucci, Mike Clark, Nancy Linford, Parag Mallick, Rafal Smigdrozki, Raven Hanna, Robert Bradbury, Ryan Grant, Sarah Black, Simon Windor, Stuart Updegrave, Ty Landrum, and doubtless many others who I'm not remembering at this moment.

Finally, I must thank my research assistant Tess Bruney who helped ensure that my claims rested on sound scientific ground, my agent Ted Weinstein whose sage advice guided me through every step of the process, and my Broadway Books editor Becky Cole (for the hardcover edition) whose critical eye and incredible enthusiasm helped make this the book it is.

References

Introduction

7 For a good analysis of Kass's history, see Chris Mooney, "Irrationalist in Chief," *The American Prospect*, vol. 12 no. 17, September 24, 2001 - October 8, 2001.

7 *"There are certain things":* Francis Fukuyama, *Our Posthuman Future*

7 *"A ban on cloning":* William Kristol, *The Future Is Now: America Confronts the New Genetics*

8 *"We need to do an unlikely thing":* McKibben, *Enough*

9 *"Regulation that flies in the face of reality":* Presidential Council on Bio-Ethics, Session 1: Pediatric Psychopharmacology, Steven E. Hyman, MD, Provost of Harvard University; Professor of Neurobiology, Harvard Medical School; and former Director, US National Institutes of Mental Health, Thursday March 6th 2003

10 *"The finitude of life is a blessing":* Leon Kass, *Life, Liberty, and the Pursuit of Human Dignity*

Chapter 1

13 For the story of Ashanti de Silva and the first gene therapy attempt on a human being, see:

W. French Anderson. Gene Therapy: The Best of Times, the Worst of Times. http://www.frenchanderson.org/french/docarticles/best.html

Results from First Human Gene Therapy Clinical Trial. http://www.hhs.gov/news/press/1995pres/951019a.html

Blaese, R. Michael, Culver, Kenneth W., Miller, A. Dusty, Carter, Charles S., Fleisher, Thomas, Clerici, Mario, Shearer, Gene, Chang, Lauren, Chiang, Yawen, Tolstoshev, Paul, Greenblatt, Jay J., Rosenberg, Steven A., Klein, Harvey, Berger, Melvin, Mullen, Craig A., Ramsey, W. Jay, Muul, Linda, Morgan, Richard A., Anderson, W. French. T Lymphocyte-Directed Gene Therapy for ADA- SCID: Initial Trial Results After 4 Years. *Science* 1995 270: 475-480.

Bordignon, Claudio, Notarangelo, Luigi D., Nobili, Nadia, Ferrari, Giuliana, Casorati, Giulia, Panina, Paola, Mazzolari, Evelina, Maggioni, Daniela, Rossi, Claudia, Servida, Paolo, Ugazio, Alberto

G., Mavilio, Fulvio. Gene Therapy in Peripheral Blood Lymphocytes and Bone Marrow for ADA- Immunodeficient Patients. *Science* 1995 270: 470-475.

13 *Cohen and Boyer:* 1996 Lemelson-MIT Prize Winners – biography of Cohen and Boyer, http://web.mit.edu/invent/a-winners/a-boyercohen.html

16 *"What choice did we have?":* Ruth Sorelle, The Gene Doctors, *Houston Chronicle*, http://www.chron.com/content/houston/interactive/special/gene/gfirst.html

17 *gene therapy blossomed:* Tim Friend, Elusive gene therapy forges on , *USA TODAY*, http://www.usatoday.com/news/health/2003-02-23-gene-therapy_x.htm

18 Anemia patient statistics are from Hanna, Leslie, Drug Watch:: Epoietin alfa (EPO) for Anemia. Source: Project Inform. *The HIV Drug Book. Second edition*. Pocket Books, New York. 1998.

18 *kidneys produce EPO*: Adamson JW. The promise of recombinant human erythropoietin. *Semin Hematol*. 1989 Apr;26(2 Suppl 2):5-8. Review.

Tripathy SK, Goldwasser E, Lu MM, Barr E, Leiden JM. Stable delivery of physiologic levels of recombinant erythropoietin to the systemic circulation by intramuscular injection of replication-defective adenovirus. *Proc Natl Acad Sci* U S A. 1994 Nov 22;91(24):11557-61.

Jerry E Bishop, Gene Therapy Method May Offer Hope in Treating Anemia With Protein EPO, *Wall Street Journal*, 11/10/93

18 The cost of EPO therapy is discussed in: John Tagliabue, Mystery Effect in Biotech Drug Puts Its Maker on Defensive. *New York Times* 2 Oct 2002

18 *first animal study of EPO gene therapy* Jeffrey M. Leiden et al., Long-term erythropoietin expression in rodents and non-human primates following intramuscular injection of a replication defective adenoviral vector, *Human Gene Therapy*, 1997, Oct 10; 8(15): p1797

18 *Monkeys went from: Gene Therapy*, vol 5, p 665

19 *Chiron and Ariad:* Victor M. Rivera, James Wilson, et al., Long Term Regulated Gene Expression in Non Human Primates, *Molecular Therapy* Vol 7, No. 5, May 2003.

20 *Team Festina:* Spalding BJ, Black-market biotechnology: athletes abuse EPO and HGH. *Biotechnology* (N Y). 1991 Nov;9(11):1050,

1052-3.

Christie Aschwanden, Gene Cheats. *New Scientist*, January 15, 2000.

Drug slush fund, Festina's doctor says team manager obtained drugs for riders, *CNN Sports Illustrated*, Tuesday July 21, 1998, http://sportsillustrated.cnn.com/cycling/1998/tourdefrance/news/1998/07/21/drugs_slush/

20 *EPO can boost your performance:* Glenn Zorpette, All Doped Up--and Going for the Gold. Miscues by the International Olympic Committee frustrate scientists developing tests for the performance-enhancing drugs erythropoietin and human growth hormone. May 21, 2000, pp. 20,22.

21 *Other athletes have used:* E. Randy Eichner. Ergogenic Aids: What Athletes Are Using--and Why. *The Physician and Sports Medicine* 25(4). Available at http://www.physsportsmed.com/issues/1997/04apr/eichner.htm

23 *Katie Binley:* Binley K, Askham Z, Iqball S, Spearman H, Martin L, de Alwis M, Thrasher AJ, Ali RR, Maxwell PH, Kingsman S, Naylor S. Long-term reversal of chronic anemia using a hypoxia-regulated erythropoietin gene therapy. *Blood*. 2002 Oct 1;100(7):2406-13.

Pollock R, Clackson T. Dimerizer-regulated gene expression. *Curr Opin Biotechnol* .13(5); 1 October 2002, pp. 459-67.

23 For an overview of ALS, see Oregon and Southwest Washington Chapter of the ALS Association web site. http://www.alsa-or.org/info/Statistics

24 Results from the IGF-1 gene therapy study are reported in: Kaspar BK, Llado J, Sherkat N, Rothstein JD, Gage FH. Retrograde viral delivery of IGF-1 prolongs survival in a mouse ALS model. *Science.* 2003 Aug 8;301(5634):839-42.

25 Reports and discussion of the use of IGF-1 gene therapy to boost mouse muscle strength (and possibly that of humans) can be found at:

Barton-Davis ER, Shoturma DI, Musaro A, Rosenthal N, Sweeney HL. Viral mediated expression of insulin-like growth factor I blocks the aging-related loss of skeletal muscle function . *Proc Natl Acad Sci* U S A 95; December 1998, 15603–15607.

Lee Sweeney testimony to President's Council on Bioethics (PCBE), Friday September 13th 2002

Musaro A, McCullagh K, Paul A, Houghton L, Dobrowolny G, Molinaro M, Barton ER, Sweeney HL, Rosenthal N. Localized Igf-1 transgene expression sustains hypertrophy and regeneration in senescent skeletal muscle. *Nat Genet.* 2001 Feb;27(2):195-200.

25 Goldspink's work is published in: Hill M, Goldspink G. Expression and splicing of the insulin-like growth factor gene in rodent muscle is associated with muscle satellite (stem) cell activation following local tissue damage. *J Physiol.* 2003 Jun 1;549(Pt 2):409-18. Epub 2003 Apr 11.

25 Se-Jin Lee's work with myostatin is described in Alexandra C. McPherron, Ann M. Lawler, and Se-Jin Lee (1997) Regulation of Skeletal Muscle Mass in Mice by a New TGF-ß Superfamily Member. *Nature* 387:83-90

25 *knocking out myostatin:* Bogdanovich, S. *et al.* Functional improvement of dystrophic muscle by myostatin blockade. *Nature* 420, 418-421 (2002).

Alexandra C. McPherron, Ann M. Lawler, and Se-Jin Lee (1997) Regulation of Skeletal Muscle Mass in Mice by a New TGF-ß Superfamily Member. *Nature* 387:83-90

26 *sarcopenia:* Evans, W. & Rosenberg, I. (1991) Biomarkers. Simon & Schuster, New York, NY.

Sarcopenia: origins and clinical relevance. *J Nutr.* 1997 May;127(5 Suppl):990S-991S. Review.

26 *"Just say you'd like your pectoralis muscle":* Sweeney PCBE testimony

26 *more than 8 million people:* Cosmetic Plastic Surgery Research: Statistics and Trends for 2001 and 2002, American Society for Aesthetic Plastic Surgery (2003)

27 *Leptin gene therapy:* Patrick Muzzin et al., Correction of obesity and diabetes in genetically obese mice by leptin gene therapy, *Proc. Natl. Acad. Sci.* USA, Vol. 93, pp. 14804–14808, December 1996

27 *"couch potatoes dream":* Eric Benjamin Lowe. UF researchers explore gene therapy to treat obesity . The Friday Evening Post: the *University of Florida Health Science Online Newsletter*; June 25, 1999. Available at http://www.health.ufl.edu/post/jun99.html

27 *epitan:* Epitan company home page http://www.epitan.com.au/

27 *albino mice whose skin can change color:* Cronin, C. A., Gluba, W. & Scrable, H., The lac operator-repressor system is functional in the mouse, *Genes and Development* 15, 1506–1517; 2001

28 *gene therapy for baldness:* Norimitsu Saito, Ming Zhao, Lingna Li, Eugene Baranov, Meng Yang, Yukinori Ohta, Kensei Katsuoka, Sheldon Penman, and Robert M. Hoffman, High efficiency genetic modification of hair follicles and growing hair shafts, *Proc. Natl.*

Acad. Sci. October 1, 2002 vol. 99 no. 20 13120-13124

29 For details on the French leukemia cases resulting from viral gene therapy, see:

Hacein-Bey-Abina S, von Kalle C, Schmidt M, Le Deist F, Wulffraat N, McIntyre E, Radford I, Villeval JL, Fraser CC, Cavazzana-Calvo M, Fischer A. A serious adverse event after successful gene therapy for X-linked severe combined immunodeficiency. *N Engl J Med.* 2003 Jan 16;348(3):255-6.

Lyford, Jo. Gene therapy 'caused T-cell leukemia'. *The Scientist.* October 20, 2003. http://www.biomedcentral.com/news/20031020/02 accessed on November 1, 2003.

LMO2-Associated Clonal T Cell Proliferation in Two Patients after Gene Therapy for SCID-X1 A. Fischer, M. Cavazzana-Calvo, *Science,* Vol 302 17 October 2003

31 *DNA nanoball gene therapy*: Sylvia Pagán Westphal, DNA nanoballs boost gene therapy, *New Scientist*, 12 May 02

32 *intron gene therapy:* Guo H, Karberg M, Long M, Jones JP III, Sullenger B, and Lambowitz AM. Group II introns designed to insert into therapeutically relevant DNA target sites in human cells. *Science* 2000;289:452-7

33 *"more like a rifle shot":* University of Texas, Austin Press Release, Cellular Mechanisms for Delivering Genetic Therapies, July 28th 2000. http://www.newswise.com/articles/2000/7/INTRON.TXA.html

33 *"bidding defiance":* A HISTORY OF THE WARFARE OF SCIENCE WITH THEOLOGY IN CHRISTENDOM, BY ANDREW DICKSON WHITE LL.D. (YALE), L.H.D. (COLUMBIA), PH.DR. (JENA) LATE PRESIDENT AND PROFESSOR OF HISTORY AT CORNELL UNIVERSITY, TWO VOLUMES COMBINED, NEW YORK D. APPLETON AND COMPANY 1898

33 *Queen Victoria*: Donald Caton, M.D., John Snow's Practice of Obstetric Anesthesia, *Anesthesiology*, 2000; 92:247–52,

33 *one tenth of one percent:* Tufts Center for the Study of Drug Development, fts Center for the Study of Drug Development Pegs Cost of a New Prescription Medicine at $802 Million, 11/30/2001, http://csdd.tufts.edu/NewsEvents/RecentNews.asp?newsid=6

35 *first artificial heart patient:* First Artificial Heart Recipient, Robert Tools, Dies at 59, November 30, 2001, By Associated Press

35 *"what's the worst that can happen to me?":* Human Genetics, Concepts and Applications, Ricki Lake, page 392

36 *$17 billion:* Entrepreneur magazine, January 2001

37 *emergency room visits:* DanceSafe, The need for adulterant screening programs, http://www.dancesafe.org/documents/druginfo/pilltesting.php

Chapter 2

39 *Statistics about Alzheimer's disease*: Alzheimer's Association web site. Available at http://www.alz.org/AboutAD/Statistics.htm

39 The NIA spending is an estimate by gerontoligist Aubrey de Grey shared during informal conversation.

39 *NGF gene therapy:* Tuszynski MH, Thal L, U HS, Pay MM, Blesch A, Conner J, Vahlsing HL. Nerve growth factor gene therapy for Alzheimer's disease. *J Mol Neurosci*. 2002 Aug-Oct;19(1-2):207.

40 *NGF gene therapy human trials:* For a review of these findings see: Tuszynski MH., Growth-factor gene therapy for neurodegenerative disorders., *Lancet Neurol*. 2002 May;1(1):51-7.

For information on ongoing clinical trials: Tuszynski MH. Gene Therapy for Alzheimer's Disease: Clinical Trial Information. Tuszynski Lab web site ;20 December 2002. http://tuszynskilab.ucsd.edu/clinical_study.htm

40 *improve the learning and memory of normal mice:* Brooks AI, Cory-Slechta DA, Federoff HJ. Gene-experience interaction alters the cholinergic septohippocampal pathway of mice. *Proc Natl Acad Sci* U S A. 2000 Nov 21;97(24):13378-83.

40 *Doogie mice:* Tsien JZ. Building a brainier mouse. *Sci Am.* 2000 Apr;282(4):62-8.

Tang YP, Shimizu E, Dube GR, Rampon C, Kerchner GA, Zhuo M, Liu G, Tsien JZ. Genetic enhancement of learning and memory in mice. *Nature*. 1999 Sep 2;401(6748):63-9.

42 *feel major pain longer:* Wei F, Wang GD, Kerchner GA, Kim SJ, Xu HM, Chen ZF, Zhuo M. Genetic enhancement of inflammatory pain by forebrain NR2B overexpression. *Nat Neurosci*. 2001 Feb;4(2):164-9.

Stull DL. Better Mouse Memory Comes at a Price. *The Scientist* 15[7]:21, Apr. 2, 2001

42 *Alzheimer's patients had lower than normal activity:* Dickey CA, Loring JF, Montgomery J, Gordon MN, Eastman PS, Morgan D , Selectively reduced expression of synaptic plasticity-related genes in amyloid precursor protein + presenilin-1 transgenic mice, *J*

Neurosci. 2003 Jun 15;23(12):5219-26

42 *CREB:* Wade, Nicholas. 3 Share Nobel Prize in Medicine for Studies of the Brain. *The New York Times*. 10 October 2000.

Kaang BK, Kandel ER, Grant SG. Activation of cAMP-responsive genes by stimuli that produce long-term facilitation in Aplysia sensory neurons. *Neuron*. 1993 Mar;10(3):427-35.

Dash PK, Hochner B, Kandel ER. Injection of the cAMP-responsive element into the nucleus of Aplysia sensory neurons blocks long-term facilitation. *Nature*. 1990 Jun 21;345(6277):718-21.

43 *fruit flies had astoundingly good memories:* Yin, J, Tully, T, CREB as a memory modulator: induced expression of a dCREB2 activator isoform enhances long-term memory in Drosophila. *Cell* 81, 107-115 (1995)

43 *could learn things in 2 trials:* Barad M, Bourtchouladze R, Winder DG, Golan H, Kandel E., Rolipram, a type IV-specific phosphodiesterase inhibitor, facilitates the establishment of long-lasting long-term potentiation and improves memory. *Proc Natl Acad Sci* U S A. 1998 Dec 8;95(25):15020-5.

Tim Tully, Rusiko Bourtchouladze, Rod Scott and John Tallman, Targeting the CREB Pathway for Memory Enhanceers, *Nature Reviews Drug Discovery*, Vol 2, April 2003, p267

43 More information about Memory Pharmaceuticals and Helicon Therapeutics is available at their web sites:

http://www.memorypharma.com/

http://www.helicontherapeutics.com/

43 The large market of people suffering from Alzheimer's or Age Associate Cognitive Impairment and its impact on the race for memory boosting drugs is documented in many places. See for example:

Sramek JJ, Veroff AE, Cutler NR. The status of ongoing trials for mild cognitive impairment. *Expert Opin Investig Drugs*. 2001 Apr;10(4):741-52.

Helmuth L. All in your mind. *Sci Aging Knowledge Environ*. 2003 Feb 26;2003(8):NS3.

Robert Langreth. Viagra for the Brain. *Forbes Magazine*, 02.04.02. http://www.forbes.com/forbes/2002/0204/046.html

President's Council on Bioethics. "Better" Memories? The Promise and Perils of Pharmacological Interventions (Staff Working Paper) http://bioethicsprint.bioethics.gov/background/better_memories.ht

ml

44 *"memory enhancers could become 'lifestyle' drugs":* Tim Tully et al, Targeting the CREB Pathway for Memory Enhancers, *Nature Reviews Drug Discovery* Volume 2, April 2003, p267

44 *"It's not an 'if' – It's a 'when'":* Robert Langreth, Viagra for the Brain, Forbes Magazine, Feb 4, 2002.

45 *caffeine and nicotine:* Mumenthaler MS, Yesavage JA, Taylor JL, O'Hara R, Friedman L, Lee H, Kraemer HC., Psychoactive drugs and pilot performance: a comparison of nicotine, donepezil, and alcohol effects., *Neuropsychopharmacology*. 2003 Jul;28(7):1366-73.

45 *15 million prescriptions:* Testimony of DEA Deputy Director Terrence Woodworth to Congress, May 16th 2000, http://www.dea.gov/pubs/cngrtest/ct051600.htm

46 *give the same mental boosts to normal children:* Rapoport JL, Dextroamphetamine. Its cognitive and behavioral effects in normal and hyperactive boys and normal men. *Arch Gen Psychiatry*. 1980 Aug:37(8):933-43

46 *"even if you have never been diagnosed":* Ken Livingston, Ritalin: Miracle Drug or Cop-Out, *The Public Interest*, No 127(Spring 1997), pp. 3-18

46 *"universal performance enhancer":* Diller comments to President's Council on Bio-Ethics

46 For a discussion of Modafinil / Provigil see:

Onion, Amanda. The No-Doze Soldier: Military Seeking Radical Ways of Stumping Need for Sleep. *ABCnews.com.*

http://abcnews.go.com/sections/scitech/DailyNews/nosleep021218.html (Accessed 23 November 2003)

Buguet A, Moroz DE, Radomski MW. Modafinil--medical considerations for use in sustained operations. *Aviat Space Environ Med*. 2003 Jun;74(6 Pt 1):659-63.

Pigeau R, Naitoh P, Buguet A, McCann C, Baranski J, Taylor M, Thompson M, MacK I I. Modafinil, d-amphetamine and placebo during 64 hours of sustained mental work. I. Effects on mood, fatigue, cognitive performance and body temperature. *J Sleep Res*. 1995 Dec;4(4):212-228.

Batejat DM, Lagarde DP. Naps and modafinil as countermeasures for the effects of sleep deprivation on cognitive performance. *Aviat Space Environ Med*. 1999 May;70(5):493-8.

Lyons TJ, French J. Modafinil: the unique properties of a new stimulant. *Aviat Space Environ Med*. 1991 May;62(5):432-5.

Buguet A, et al, Modafinil, d-amphetamine and placebo during 64 hours of sustained mental work. Effects on two nights of recovery sleep. *Journal of Sleep Research*, 1995 Dec;4(4):229-241

47 Some of the effects of Paxil and other SSRI's on non-depressed humans are reported at:

Knutson B, Wolkowitz OM, Cole SW, Chan T, Moore EA, Johnson RC, Terpestra J, Turner RA, Reus VI. Selective alteration of personality and social behavior by serotonergic intervention. *American Journal of Psychiatry* 1998; 155: 373-379

Verkes RJ, Van der Mast RC, Hengeveld MW, Twyl JP, Zwinderman AH, Van Kempen EM. Reduction by paroxetine of suicidal behavior in patients with repeated suicide attempts but not major depression. *American Journal of Psychiatry* 1998; 155: 543-547

The link between serotonin and social dominance in primates has been reported in numerous studies, including:

Raleigh MJ, Brammer GL, McGuire MT, Yuwiler A. Dominant social status facilitates the behavioral effects of serotonergic agonists. *Brain Res*. 1985 Dec 2;348(2):274-82.

Raleigh MJ, Brammer GL, McGuire MT. Male dominance, serotonergic systems, and the behavioral and physiological effects of drugs in vervet monkeys (Cercopithecus aethiops sabaeus). *Prog Clin Biol* Res. 1983;131:185-97

47 Research into genetic contributions to ADHD is summarizes in OMIM entry 143465 ATTENTION DEFICIT-HYPERACTIVITY DISORDER; ADHD

48 *bipoloar disorder genes:* Hattori E, Liu C, Badner JA, Bonner TI, Christian SL, Maheshwari M, Detera-Wadleigh SD, Gibbs RA, Gershon ES, Polymorphisms at the G72/G30 gene locus, on 13q33, are associated with bipolar disorder in two independent pedigree series., *Am J Hum Genet*. 2003 May;72(5):1131-40. Epub 2003 Mar 19

48 *mutations in a gene for the serotonin transporters* Lesch KP, Gross J, Franzek E, Wolozin BL, Riederer P, Murphy DL., Primary structure of the serotonin transporter in unipolar depression and bipolar disorder., *Biol Psychiatry*. 1995 Feb 15;37(4):215-23

48 For a summary of research in the area of gene therapy for pain control, see:

Fink D, Mata M, Glorioso JC., Cell and gene therapy in the treatment of pain., *Adv Drug Deliv Rev*. 2003 Aug 28;55(8):1055-64

Cancer Pain Control Possible with Gene Therapy, October 15th 2002, University of Pittsburgh Medical Center press release

48 *cut their drinking habits:* Thanos PK, Volkow ND, Freimuth P, Umegaki H, Ikari H, Roth G, Ingram DK, Hitzemann R., Overexpression of dopamine D2 receptors reduces alcohol self-administration., *J Neurochem*. 2001 Sep;78(5):1094-103

49 *prairie voles:* Pitkow LJ, Sharer CA, Ren X, Insel TR, Terwilliger EF, Young LJ., Facilitation of affiliation and pair-bond formation by vasopressin receptor gene transfer into the ventral forebrain of a monogamous vole., *J Neurosci*. 2001 Sep 15;21(18):7392-6.

49 The link between variants of the 5-HT2 receptor and psychedelic and spiritual experiences is documented in several locations, including:

LeDOUX J. The Self: Clues from the Brain. *Ann N Y Acad Sci*. 2003 Oct;1001:295-304.

Lauterbach EC, Abdelhamid A, Annandale JB. Posthallucinogen-like visual illusions (palinopsia) with risperidone in a patient without previous hallucinogen exposure: possible relation to serotonin 5HT2a receptor blockade. *Pharmacopsychiatry*. 2000 Jan;33(1):38-41.

Kurup RK, Kurup PA. Hypothalamic digoxin, hemispheric chemical dominance, and spirituality. *Int J Neurosci*. 2003 Mar;113(3):383-93.

Dean Hamer's research into a specific genetic link to male homosexuality is still up in the air. However it's clear from population genetic that there is some linkage, just not whether the chromosome region Hamer identified contributes to it. Hamer's original paper is at:

Hamer, D. H., Hu, S., Magnuson, V. L., Hu, N. & Pattatucci, A. M. L. (1993). A linkage between DNA markers on the X chromosome and male sexual orientation. *Science* 261, 321–7.

A discussion of the genetic linkage to religion can be found at:

Bouchard TJ Jr, McGue M, Lykken D, Tellegen A. Intrinsic and extrinsic religiousness: genetic and environmental influences and personality correlates. *Twin Res*. 1999 Jun;2(2):88-98.

Karen Carver and J. Richard Udry. The Biosocial Transmission of Religious Attitudes. *The National Longitudinal Study of Adolescent Health*. Presentation 1997. Abstract available online at http://www.cpc.unc.edu/projects/addhealth/abs2.html

50 *$300 billion per year:* Testimony of James H. Scully, Jr., M.D.

American Psychiatric Association

August 15, 1996, before the Pennsylvania State House Insurance and Health and Human Services Committees on the Mental Health Parity Act House Bill 2237, PN 1872

50 *reducing the incidence of Alzheimer's:* Citizens for Long Term Care, Long Term Care: An Overview, http://citizensforltc.org/issues_overview.html

51 *correlation between average IQ and rate of economic growth:* Intelligence and the Wealth and Poverty of Nations, RICHARD LYNN, University of Ulster, Coleraine, Northern Ireland, TATU VANHANEN, University of Helsinki, Finland (2002) http://www.rlynn.co.uk/pages/articles.htm

51 *every additional year of education:* Losing out - how UK higher education is losing vital investment, Association of University Teachers, June 2000 Research Report, http://www.aut.org.uk/media/html/losingvitalinvestment1.html

52 *attitudes towards genetic engineering:* Public Awareness and Attitudes about Reproductive Genetic Technology, The Genetics and Public Policy Center with Princeton Survey Research Associates, December 9, 2002

Chapter 3

56 *close to a billion dollars:* Tufts Center for the Study of Drug Development, Tufts Center for the Study of Drug Development Pegs Cost of a New Prescription Medicine at $802 Million, 11/30/2001, http://csdd.tufts.edu/NewsEvents/RecentNews.asp?newsid=6

56 *drugs whose patents have expired:* HOW INCREASED COMPETITION FROM GENERIC DRUGS HAS AFFECTED PRICES AND RETURNS IN THE PHARMACEUTICAL INDUSTRY, The Congress of the United States, Congressional Budget Office, JULY 1998

60 *life expectancy vs. GDP:* Data courtesy of World Resources Institute and UN Human Development Report 2003

68 *$6,911 per student:* Adapted from "Revenues and Expenditures for Public Elementary and Secondary Education: School Year 1999-2000" by Frank Johnson, U.S. Department of Education, National Center for Education Statistics, May 16, 2002

Chapter 4

69 *In 1900 a man:* Hayflick L. How and why we age. *Exp Gerontol.* 1998 Nov-Dec; 33(7-8): 639-53. Review.

70 *"if, indeed, we were to find a cure for cancer":* The President's Council on Bioethics. Duration of Life: Is There a Biological Warranty Period? Testimony given on 12 December 2002 to the PCB by S. Jay Olshansky. Available at http://www.bioethics.gov/transcripts/dec02/session2.html

71 *$3.6 Billion:* Entrepreneur magazine, January 2001

72 *position paper on human aging:* S. Jay Olshansky, Leonard Hayflick and Bruce A. Carnes, The Truth about Human Aging, *Scientific American*, May 13th 2002,

72 *"in the past few years":* Lithgow comment at Anti-Aging Drug Discovery Summit, October 2002, San Francisco CA

72 *Andrzej Bartke:* Bartke et al, Extending the lifespan of long-lived mice, *Nature*, VOL 414, 22 NOVEMBER 2001

73 *a single gene could double the lifespan:* Johnson TE. Increased life span of age-1 mutants in Caenorhabditis elegans and lower Gompertz rate of aging. *Science.* 1990;249:908-912

73 *"worms have 100 million nucleotides":* Steve Austad testimony to PCBE, December 12th 2002, http://www.bioethics.gov/december02/session1.html

73 *"my initial findings":* Email interview with Tom Johson

74 A comprehensive database of genetic changes which lengthen animal lifespan can be find at Science Magazine's SAGE KE. http://sageke.sciencemag.org/

75 *more resistant to the external stress:* Lithgow GJ, White TM, Melov S, Johnson TE. Thermotolerance and extended life-span conferred by single-gene mutations and induced by thermal stress. *Proc Natl Acad Sci* U S A. 1995 Aug 1;92(16):7540-4

75 *added resistance to stresses:* Oxidants, oxidative stress and the biology of ageing, Toren Finkel & Nikki J. Holbrook, *NATURE* VOL 408, 9 NOVEMBER 2000

76 *how they work in mice:* H. R. Warner, Subfield History: Use of Model Organisms in the Search for Human Aging Genes. *Science's SAGE KE* (12 February 2003)

77 *Holzenberger's mice:* M. Holzenberger et al, IGF-1 receptor regulates lifespan and resistance to oxidative stress in mice. *Nature*, 4 December 2002 [e-pub ahead of print]

77 *Researchers at Harvard:* M. Bluher, B. B. Kahn, R. C. Kahn, Extended Longevity in Mice Lacking the Insulin Receptor in Adipose Tissue, *Science* 299 572 (2003)

78 *"It's wonderful":* R. J. Davenport, One for All. *Science's SAGE KE* (11 December 2002), http://sageke.sciencemag.org/cgi/content/full/sageke;2002/49/nf15

80 There are many reviews of the free radical theory of aging. For example, see

Biesalski HK. Free radical theory of aging. *Curr Opin Clin Nutr Metab Care*. 2002 Jan;5(1):5-10. Review

Or

Kowald A. The mitochondrial theory of aging. *Biol Signals Recept*. 2001 May-Aug;10(3-4):162-75. Review.

81 *produce more SOD:* Sun J, Folk D, Bradley TJ, Tower J. Induced overexpression of mitochondrial Mn-superoxide dismutase extends the life span of adult Drosophila melanogaster. *Genetics*. 2002 Jun;161(2):661-72.

81 *methuselah mice:* Spencer CC, Howell CE, Wright AR, Promislow DE. Testing an 'aging gene' in long-lived drosophila strains: increased longevity depends on sex and genetic background. *Aging Cell.* 2003 Apr;2(2):123-30.

82 *p66shc seems to make the cells:* Migliaccio E, Giorgio M, Mele S, Pelicci G, Reboldi P, Pandolfi PP, Lanfrancone L, Pelicci PG. The p66shc adaptor protein controls oxidative stress response and life span in mammals. *Nature*. 1999 Nov 18;402(6759):309-13.

82 *EUK-8 and EUK-134:* Melov S, Ravenscroft J, Malik S, Gill MS, Walker DW, Clayton PE, Wallace DC, Malfroy B, Doctrow SR, Lithgow GJ. Extension of life-span with superoxide dismutase/catalase mimetics. *Science*. 2000 Sep 1;289(5484):1567-9.

Chapter 5

83 *caloric restriction:* Weindruch R, Walford RL. The retardation of aging and disease by dietary restriction. Springfield, Ill.: Charles C Thomas, 1988.

83 *known for almost a century:* Hursting SD, Lavigne JA, Berrigan D, Perkins SN, Barrett JC. Calorie restriction, aging, and cancer prevention: mechanisms of action and applicability to humans. *Annu Rev Med*. 2003; 54: 131-52. Epub 2001 Dec 03.

84 *"one of the striking things":* Steve Austad testimony to PCBE, December 12th 2002, http://www.bioethics.gov/december02/session1.html

85 *slows the rate of death of neurons:* UF STUDY: CALORIE RESTRICTION REDUCES AGE-RELATED BRAIN CELL DEATH, Dec. 30, 2002, http://www.napa.ufl.edu/2002news/localoriebrains.htm

85 *biochemistry of younger animals:* Roth GS, Lane MA, Ingram DK, Mattison JA, Elahi D, Tobin JD, Muller D, Metter EJ. Biomarkers of caloric restriction may predict longevity in humans. *Science*. 2002 Aug 2;297(5582):811.

85 *compressed morbidity:* Masoro EJ. Subfield history: caloric restriction, slowing aging, and extending life. *Sci Aging Knowledge Environ*. 2003 Feb 26;2003(8):RE2.

86 *caloric restriction on a colony of rhesus monkeys:* Roth GS, Ingram DK, Lane MA. Caloric restriction in primates and relevance to humans. *Ann N Y Acad Sci*. 2001 Apr;928:305-15

87 *caloric restriction side effects:* Hopkin K. Dietary drawbacks. *Sci Aging Knowledge Environ*. 2003 Feb 26;2003(8):NS4.

88 *INDY gene:* Blanka Rogina, Robert A. Reenan, Steven P. Nilsen, Stephen L. Helfand, Extended Lifespan Conferred by Cotransporter Gene Mutations in *Drosophila*, *Science*, 290 2137-2140 (2000)

88 *Rpd3 lived between 33 percent and 50 percent longer:* Blanka Rogina, Stephen L. Helfand, Stewart Frankel, Longevity Regulation by *Drosophila* Rpd3 Deacetylase and Caloric Restriction, *Science* 298, 1745 (2002)

88 *4-phenyl butyrate extended their lives:* Hyung-Lyun Kang, Seymour Benzer, and Kyung-Tai Min, Life extension in Drosophila by feeding a drug, *Proceedings of the National Academy of Science*, vol 99, no. 2, 838-843

89 *caloric restriction mimetics:* Lane MA, Ingram DK, Roth GS. The serious search for an anti-aging pill. *Sci Am.* 2002 Aug;287(2):36-41.

89 *genes that are expressed differently:* Biomarker Pharmaceuticals, Inc. Technology [web site]. Accessed 2/2/2004. Available at http://www.biomarkerinc.com/html/technology.htm

90 *all turn out to be related:* Gat-Yablonski G, Ben-Ari T, Shtaif B, Potievsky O, Moran O, Eshet R, Maor G, Segev Y, Phillip M. Leptin reverses the inhibitory effect of caloric restriction on longitudinal growth. *Endocrinology*. 2004 Jan;145(1):343-50. Epub 2003 Oct 02.

90 *preserve healthy youthful life:* Lee CK, Klopp RG, Weindruch R, Prolla TA. Gene expression profile of aging and its retardation by caloric restriction. *Science*. 1999 Aug 27;285(5432):1390-3.

Sohal RS, Weindruch R. Oxidative stress, caloric restriction, and aging. *Science*. 1996 Jul 5;273(5271):59-63.

91 *75% life extension:* M. Beckman, Dieting Dwarves Live It Up. *Sci.*

SAGE KE 2001 (8), nf4 (2001).

92 *"indefinite postponement of aging may be within sight":* de Grey AD, Ames BN, Andersen JK, Bartke A, Campisi J, Heward CB, McCarter RJ, Stock G. Time to talk SENS: critiquing the immutability of human aging. *Ann N Y Acad Sci*. 2002 Apr;959:452-62; discussion 463-5.

93 *succeeded in moving a mitochondrial gene:* Yeast mitochondrial ATPase subunit 8, normally a mitochondrial gene product, expressed in vitro and imported back into the organelle, DP Gearing, P Nagley, *EMBO J*. 1986 Dec 20;5(13):3651-5

95 *"the lights are green everywhere you go":* Kolata, G., Pushing limits of the human life span. *New York Times*, 9 March 1999.

95 *"Aging is not immutable.":* Aging Research Grows Up

95 *"I think it's easily possible":* Austad testimony to PCBE

95 *effective life-extension drugs on the market:* Thomas E. Johnson, Subfield History: *Caenorhabditis elegance* as a System for Analysis of the Genetics of Aging, *Sage KE* August 28th 2002, http://sageke.sciencemag.org/cgi/content/full/sageke;2002/34/re4

Chapter 6

98 Data for this graphic is consolidated from a number of sources:

John R. Wilmoth, Chapter 2: Human Longevity in Historical Perspective, World Population Prospects 2000, United Nations Population Division

Economic Growth and the State of Humanity, by Indur M. Goklany. PERC Policy Series PS-21, Bozeman, MT: PERC, April 2001.

Getting better all the time, Nov 8th 2001, *The Economist*, http://economist.com/surveys/displaystory.cfm?story_id=841842

99 *Less Developed Nation Life Expectancy:* World Population Prospects 2000, UN Population Division.

99 *one in five Americans:* National Center for Chronic Disease Prevention and Health Promotion, *Healthy Aging for Older Adults*

100 *average citizen of Japan:* UN World Population Prospects, 2002 Revision

100 *spend anywhere from three to five times:* Howard Oxley, Stephane Jacobzone, *Health Care Expenditures: A Future in Question* (2001)

101 *almost half of all health care spending:* The Demographic Facts of Life, http://www.populationconnection.org/Communications/demfacts.PD

F

101 *15 percent of US GDP in 2030:* Congressional Budget Office, A 125-Year Picture of the Federal Government's Share of the Economy, 1950 to 2075

102 Medicare costs in last year are taken from

"Medical expenditures during the last year of life: Findings from the 1992-1996 Medicare current beneficiary survey," by Donald R. Hoover, Ph.D., M.P.H., Dr. Crystal, Rizie Kumar, M.S., and others, in the December 2002 Health Services Research 37(6), pp. 1625-1642

Estimates of cost in the last three years are based on simple extrapolation of the data in

Differences in Medicare Expenditures During the Last 3 Years of Life, Lisa R. Shugarman, PhD, Diane E. Campbell, PhD, Chloe E. Bird, PhD, Jon Gabel, MA, Thomas A. Louis, PhD, Joanne Lynn, MD, MS, J GEN INTERN MED 2004;19:127–135.

103 Population data is from:

Population Division of the Department of Economic and Social Affairs of the United Nations Secretariat, World Population Prospects: The 2002 Revision and World Urbanization Prospects: The 2001 Revision, http://esa.un.org/unpp, 26 March 2004; 11:23:17 AM.

105 *"The bottom line is":* Jay Olshansky testimony to PCBE

106 *9.4 billion people:* This is a simple application of the model described in the text to statistics from:

Population Division of the Department of Economic and Social Affairs of the United Nations Secretariat, World Population Prospects: The 2002 Revision and World Urbanization Prospects: The 2001 Revision, http://esa.un.org/unpp, 26 March 2004; 11:23:17 AM.

106 *between 1970 and 1973:* U.S. Bureau of the Census, International Data Base. Note: Data updated 7-17-2003

108 *the older you are the more likely you are to vote:* US Census Bureau - http://www.census.gov/population/socdemo/voting/p20-542/tab08.xls

Chapter 7

111 The story of Louise Joy Brown and the events leading up to her conception and birth is documented in many places. See:

Gwynne P, Collings A, Gastel B., The test-tube baby. *Newsweek.* 1978 Jul 24;92(4):76

Cohn V., Test-tube baby pioneer urges easing of curbs, *Washington*

Post. 1978 Oct 12;:A26

Weintraub RM., First test-tube baby born in British hospital. *Washington Post*. 1978 Jul 26;:A1

By-passing a block to conception. *Times (Lond)*. 1978 Jul 27;:15.

Washington Post. 1978 Jul 27;:A1+. Test tube baby 'well': doctors predict more successes. Nossiter B.

112 For Kass and Rifkin's opposition to IVF, see:

"to whom, in this new era, would we entrust the authority to decide what is a good gene that should be added to the gene pool and what is a bad gene that should be eliminated?" Rifkin, 1998 *The Biotech Century*.

Chris Mooney, Irrationalist in Chief., *The American Prospect*, Issue Date: 9.24.01. http://www.prospect.org/print/V12/17/mooney-c.html

Kass, Leon. Babies by Means of in vitro Fertilization: Unethical Experiments on the Unborn?, *New England Journal of Medicine* 285:1174-1179, 1971.

112 *more than a million babies:* Albert Lasker Award for Clinical Medical Research, 2001, Comments at the Awards Ceremony, Presented by Joseph L. Goldstein

112 *one out of every one hundred births:* NIRA Policy Research 2001 Vol.14 No. 6, The Development of Life Sciences and Law, http://www.nira.go.jp/publ/seiken/ev14n06.html

112 *In Australia:* IVF Defects, Catalyst, Australian Broadcasting Corporation, Thursday, 17 July 2003

112 *Doctors take several eggs:* Chapter 25a. Test Tube Babies - IVF & GIFT. from the book How to Have a Baby: Overcoming Infertility. by Dr. Aniruddha Malpani, MD and Dr. Anjali Malpani, MD. http://www.infertilitybooks.com/onlinebooks/malpani/chapter25a.html.

112 Additional information on the cost of IVF can be found at:

The Fertility Race. Twenty Years of Test-Tube Babies. A Revolutionary Birth. © Copyright 1998, Minnesota Public Radio. http://news.mpr.org/features/199711/20_smiths_fertility/part7/section1.shtml

The Cost of IVF. The National Fertility Directory. http://www.fertilitydirectory.org/costofivf.html

Dr. Sherman J. Silber. Millenium Update: Future Trends in Human Infertility and Infertility Treatment.

http://www.infertile.com/treatmnt/treats/millnium.htm

112 *no guarantees that a mother will get pregnant:* Exactly how many embryos are flushed out of the body without implanting in the uterine wall is a matter of some debate. Nevetheless, it's clear that the percentage is substantial. For more discussion, see:

Genbacev OD, Prakobphol A, Foulk RA, Krtolica AR, Ilic D, Singer MS, Yang ZQ, Kiessling LL, Rosen SD, Fisher SJ. Trophoblast L-selectin-mediated adhesion at the maternal-fetal interface. *Science*. 2003 Jan 17;299(5605):405-8.

Hull MGR, Glazener CMA, Kelly NJ, et al. Population study of causes, treatment, and outcome of infertility. *BMJ* 1985; 291: 1693-1697.

Templeton AA, Penney GC. The incidence, characteristics, and prognosis of patients whose infertility is unexplained. *Fertil Steril* 1982; 37: 175-182.

Barnea ER, Holford TR, McInnes DRA. Long-term prognosis of infertile couples with normal basic investigations: a life-table analysis. *Obstet Gynecol* 1985; 66: 24-26.

Rousseau S, Lord J, Lepage Y, Van Campenhout J. The expectancy of pregnancy for "normal" infertile couples. *Fertil Steril* 1983; 40: 768-772.

113 *one in every six couples is infertile:* p. 83 Silver, Lee. *Remaking Eden.* Avon: New York, 1997.

113 *in PGD:* Dr. Aniruddha Malpani, MD and Dr. Anjali Malpani, MD. Chapter 26: PGD - Preimplantation Genetic Diagnosis – The Newest ART. from the book How to Have a Baby: Overcoming Infertility. http://www.infertilitybooks.com/onlinebooks/malpani/chapter26.html

114 *PGD for Huntington's disease*:

Sermon K, De Rijcke M, Lissens W, De Vos A, Platteau P, Bonduelle M, Devroey P, Van Steirteghem A, Liebaers I. Preimplantation genetic diagnosis for Huntington's disease with exclusion testing. *Eur J Hum Genet*. 2002 Oct;10(10):591-8.

Lashwood A, Flinter F. Clinical and counselling implications of preimplantation genetic diagnosis for Huntington's disease in the UK. *Hum Fertil (Camb).* 2001;4(4):235-8. Review..

114 *Thousands of PGD cycles done*:

International Working Group on Preimplantation Genetics. Preimplantation Genetic Diagnosis: Experience Of Three Thousand Clinical Cycles. Report of the 11th Annual Meeting of International Working Group on Preimplantation Genetics, in Association with

10th International Congress of Human Genetics, Vienna, May 15, 2001. preimplantationgenetics.org. http://216.242.209.125/11m.shtml

Kuliev A, Verlinsky Y. Current features of preimplantation genetic diagnosis. *Reprod Biomed Online*. 2002 Nov-Dec;5(3):294-9.

115 the story of the Dunthornes is documented in "Superhuman – The Baby Makers" and is also described in:

Science screens out defective genes, *BBC Health Online*, Saturday, 18 November, 2000, 00:07 GMT

115 For information on APOE and how variants of it affect heart disease and Alzheimer's disease risk, see:

JD. Apolipoproteins and aging: emerging mechanisms. *Ageing Res Rev*. 2002 Jun;1(3):345-65.

116 *ANDI:* Chan AW, Chong KY, Martinovich C, Simerly C, Schatten G. Transgenic monkeys produced by retroviral gene transfer into mature oocytes. *Science*. 2001 Jan 12;291(5502):309-12.

117 For information on Tay Sachs, see

http://www.marchofdimes.com/professionals/681_1227.asp

118 Cost utility of prenatal diagnosis and the risk-based threshold, Ryan A Harris, A Eugene Washington, Robert F Nease Jr, Miriam Kuppermann, *Lancet* 2004; 363: 276-82

119 *73% approved:* Americans Deeply Divided About Use of Genetic Technologies in Reproduction, Genetics and Public Policy Center, Johns Hopkins University, 2002, http://www.dnapolicy.org/research/reproductiveGenetics.jhtml?subSection=reproductive

Chapter 8

121 The race to sequence the human genome is documented in many places:

Venter JC, et al., The sequence of the human genome. *Science*. 2001 Feb 16;291(5507):1304-51. Erratum in: *Science* 2001 Jun 5;292(5523):1838.

Pennisi E. Academic sequencers challenge Celera in a sprint to the finish. *Science*. 1999 Mar 19;283(5409):1822-3.

Firm's patent bids alarm gene project researchers. (the case of Celera Genomics) Paul Jacobs, Peter G. Gosselin. *Los Angeles Times* Oct 24, 1999 pA1 col 6 (35 col in)

Celera Genomics Completes Sequencing Phase Of Drosophila

Genome Project; Sequencing of Human Genome Begins. *Business Wire* Sept 9, 1999 p1104

Media hype rivalry between genome groups. (News) Marilynn Larkin. *The Lancet* August 7, 1999 v354 i9177 p496 Mag.Coll.: 100F3564.

ELIOT MARSHALL.A High-Stakes Gamble on Genome Sequencing. (Celera Genomics). *Science* June 18, 1999 v284 i5422 p1906

ELIZABETH PENNISI. Academic Sequencers Challenge Celera in a Sprint to the Finish. (Human Genome Project; Celera Genomics). *Science* March 19, 1999 v283 i5409 p1822(1)

Robert Langreth. U.S., private gene-sequencing race heats up. (Human Genome Project; Celera Genomics). *The Wall Street Journal* March 16, 1999 pB5(E) col 4 (13 col in)

The gene race quickens. (human genome mapping by Celera Genomics Corp.)(Editorial) *The Washington Post* August 22, 1998 v121 n234 pA18 col 1 (15 col in)

Favello A, Hillier L, Wilson RK. Genomic DNA sequencing methods. *Methods Cell Biol*. 1995;48:551-69.

122 *"Starting in 1990":* NHGRI workshop summary, *Sequencing and Re-sequencing the Biome!*, http://www.genome.gov/10005564, July 23rd 2002

122 *exponential improvement:* Dawkins, Richard. "Son of Moore's Law". In Brockman, John (ed.)*The Next Fifty Years*. New York: Vintage, 2002

124 *your genome is spelled out:* Human Genome Program, U.S. Department of Energy, Genomics and Its Impact on Science and Society: A 2003 Primer, 2003.

124 *gene-chips that detect at least one hundred thousand:*

BRANCA, MALORYE. Beyond the blueprint. *Bio-IT World*. April 15, 2003.

Marilee Ogren . Whole-Genome SNP Genotyping: Four highly parallel approaches make the technically intractable a reality. *The Scientist*. 17(11) Jun. 2, 2003.

126 *a couple goes into an infertility lab:* pp. 233-237, Lee Silver, *Remaking Eden*.

127 *only 20 percent:* Public Awareness and Attitudes about Reproductive Genetic Technology, The Genetics and Public Policy Center with Princeton Survey Research Associates, December 9, 2002

128 *genes for obesity:* The Human Obesity Gene Map: The 2003 Update Snyder et al., available online at http://obesitygene.pbrc.edu/cgi-

bin/ace/mainMenu.cgi

130 *"fist step towards a eugenic world":* How One Clone Leads to Another, New York Times, January 24th 2003, Leon Kass

131 *"The person left without any choice":* Bill McKibben, *Enough*

132 *MAOA:* Caspi A, McClay J, Moffitt TE, Mill J, Martin J, Craig IW, Taylor A, Poulto R. Role of genotype in the cycle of violence in maltreated children. *Science*. 2002 Aug 2;297(5582):851-4.

134 *genes and environment:* Plomin, R., DeFries, J. C., McClearn, G. E. & McGuffin, P. (2000). *Behavioral Genetics*. 4th ed. New York: Worth.

134 *genes and personality:*

Riemann, R., Angleitner, A. & Strelau, J. (1997). Genetic and environmental influences on personality: a study of twins reared together using the self- and peer-report NEO-FFI scales. *J. Pers*. 65, 449–76.

Bouchard, T. J. Jr. & Loehlin, J. C. (2001). Genes, Evolution, and Personality. Behav. *Genet*. 31, 243–73.

135 Einstein's IQ is certainly just a guess. It serves here as an illustrative example and nothing more. For more on Einstein's mental faculties, see:

Witelson SF, Kigar DL, Harvey T. The exceptional brain of Albert Einstein. *Lancet* 353(9170):2149-53, 16 June 1999.

137 *"most traits are desirable at intermediate values":* Pinker, Stephen. The Designer Baby Myth. *Boston Globe*, June 1st 2003.

137 *COMT gene and IQ* OMIM entry at http://www.ncbi.nlm.nih.gov/htbin-post/Omim/dispmim?116790

138 *families who had lost a child:* Margaret Talbot. A Desire to Duplicate. *The New York Times Magazine*, February 4, 2001. (http://www.newamerica.net/index.cfm?pg=article&pubID=95)

139 *"psychological implications"* Jeremy Rifkin and T. *Howard, Who Shall Play God?,* New York: Dell, 1977, p 15.

139 *"profoundly dehumanizing":* Kass, *Life, Liberty, and the Defense of Dignity*, page 131

140 *psychological problems:* Van Balen, F. Child-Rearing Following In Vitro Fertilisation, *Journal of Child Psychology and Psychiatry*, 37, 687-693 (1996)

140 *fantasy child:* Burns, L.H., Infertility as Boundary Ambiguity: One Theoretical Perspective, *Family Processes* 26, 359-372 (1987)

140 *perfectly normal and well adjusted:* Emma Goodman, Fiona MacCallum and Susan Golombok, Follow-up studies ion the psychological consequences of successful IVF treatment, *Biomedical Ethics*, Vol 3, 1998, No 2

141 *"the notion of enhancement is deeply ingrained":* Research offers new ways to study, alter brain function but benefits come with social policy questions, Lisa M. Krieger, *San Jose Mercury News*, Tue, May. 21, 2002

141 *increasing average height:* J. Lawrence Angel, "Colonial to Modern Skeletal Change in the U. S. A.," American *Journal of Physical Anthropology* 45, no.3 (November, 1976): 725.

141 *Flynn effect: James* R. Flynn 'The mean IQ of Americans: Massive gains 1932 to 1978'. *Psychological Bulletin*, 95, 1984, 29-51.

James R. Flynn 'Massive IQ gains in 14 nations: what IQ tests really mean'. *Psychological Bulletin*, 101, 1987, 171-91.

143 *"the impending practice of germline genetic manipulation":* Governing the Genome, Ralph Brave, *The Nation*, November 21, 2001

143 *"Modern genetics is eugenics":* The Roots of Racism and Abortion, An Exploration of Eugenics, by John Cavanaugh-O'Keefe, http://www.eugenics-watch.com/roots/index.html

143 *"our present condition is this":* Designing Our Descendants, Gilbert Meilaender, *First Things* 109 (January 2001): 25-28.

145 *two thirds of the couples:* Pre-implantation Genetic Diagnosis (PGD)-Regulation and Medical Practice -L. Hennen, A. Sauter, EPTA Annual Conference, "Research Involving Human Beings", Bern, 2003. Büro für Technikfolgen-Abschätzung beim Deutschen Bundestag

Chapter 9

149 *Johnny Ray* Me, Myself, My Implants, My Micro-processors, and I, *Software Development Magazine*, September 2000

149 Statistics on stroke frequency and number of locked in and paralyzed patients can be found at:

American Stroke Association. About Stroke (web site). http://www.strokeassociation.org/presenter.jhtml?identifier=11402

The National SCI Statistical Center. Spinal Cord Injury: Facts and figures at a glance. Birmingham, AL: University of Alabama at Birmingham Press, May 2001.

Nell Boyce. Enter the Cyborgs. *U.S. News and World Report*, May 13, 2002: 56-8.

149 *Phil Kennedy*: Interview with Phil Kennedy 2002 NIH Neural Prosthetics Workshop outside Washington D.C.

150 *Bakay implanted the device:* Kennedy PR, Bakay RA, Moore MM, Adams K, Goldwaithe J. Direct control of a computer from the human central nervous system. *IEEE Trans Rehabil Eng*. 2000 Jun;8(2):198-202

153 *brain computer interfaces:* Nicolelis M. Brain-machine interfaces to restore motor function and probe neural circuits. *Nature Reviews Neuroscience* Vol. 4, May 2003: 417-22.

153 *Nicolelis:* Mind Over Machine, Carl Zimmer, *Popular Science* magazine, February 2004

Nicolelis bio at http://www.nicolelislab.net/

154 *rat could control a robot arm:* Chapin JK, Moxon KA, Markowitz RS, Nicolelis MA. Real-time control of a robot arm using simultaneously recorded neurons in the motor cortex. *Nat Neurosci*. 1999 Jul;2(7):664-70

Wessberg J, Stambaugh CR, Kralik JD, Beck PD, Laubach M, Chapin JK, Biggs SJ, Srinivasan MA, Nicolelis MAL. Real-time prediction of hand trajectory by ensembles of cortical neurons in primates. *Nature* V408 (16 Nov 2000):361-65

154 *rhesus monkey:* Carmena JM, Lebedev MA, Crist RE, O'Doherty JE, Santucci DM, Dimitrov D, Patil PG, Henriquez CS, Nicolelis MA. Learning to Control a Brain-Machine Interface for Reaching and Grasping by Primates. *PLoS Biol.* 2003 Nov;1(2):E42. Epub 2003 Oct 13.

154 *"the monkey suddenly realized":* DukeMed News Release, Monkeys Consciously Control a Robot Arm Using Only Brain Signals; Appear to "Assimilate" Arm As If it Were Their Own, 10/13/2003

156 *tested a version of their system on humans:* Thought-Controlled Arm May Work in People-Report, *Reuters Science*, Tue Mar 23, 4:59 PM By Maggie Fox

156 *first cochlear implants:* Berliner KI, House WF. Cochlear implants: an overview and bibliography. *Am J Otol*. 1981 Jan;2(3):277-82. Review.

157 *20 million people:* Blackwell DL, Collins JG, Coles R. Summary health statistics for U.S. adults: National Health Interview Survey, 1997. *National Center for Health Statistics. Vital Health Stat* 10(205). 2002.

157 *cochlear implant:* National Institutes of Health Data, http://www.nidcd.nih.gov/health/hearing/coch.asp#c

157 *30,000 fibers:* Robert Finn, A. James Hudspeth, Jozef Zwislocki, Eric Young, and Michael Merzenich. Sound from silence: the development of cochlear implants. *National Academy of Sciences: Washington*. Available at http://www2.nas.edu/bsi. (1998)

158 *auditory cortex:* Rousche PJ, Normann RA. Chronic intracortical microstimulation (ICMS) of cat sensory cortex using the Utah Intracortical Electrode Array. *IEEE Trans Rehabil Eng*. 1999 Mar;7(1):56-68.

158 *Jens:* Steven Kotler. Vision Quest. Wired 10(9). http://www.wired.com/wired/archive/10.09/vision.html

159 *abilities beyond those of the human eye:* Dobelle WH. Artificial vision for the blind by connecting a television camera to the visual cortex. *ASAIO J*. 2000 Jan-Feb;46(1):3-9.

161 *see what the cat is looking at:* Reconstruction of Natural Scenes from Ensemble Responses in the Lateral Geniculate Nucleus, Garrett B. Stanley, Fei F. Li, and Yang Dan, *The Journal of Neuroscience*, September 15, 1999, 19(18):8036–8042

162 *communicating sounds directly from brain to brain:* Hoag H. Neuroengineering: Remote control. *Nature* 423, 796 - 798 (19 June 2003).

162 *sensation equals recollection equals imagination:* Stephen M. Kosslyn, Giorgio Ganis, William L. Thompson, Neural Foundations of Imagery, *Nature Reviews Neuroscience* 2, 635-642 (2001); doi:10.1038/35090055

163 *no cure for Parkinson's:* Tuite P, Riss J. Recent developments in the pharmacological treatment of Parkinson's disease. *Expert Opin Investig Drugs*. 2003 Aug;12(8):1335-52.

163 *Parksinson's tremors stopped:* Pollak P, Benabid AL, Limousin P, Krack P. Treatment of Parkinson's disease. New surgical treatment strategies. *Eur Neurol*. 1996;36(6):400-4.

164 Recent developments with deep brain stimulation are documented in:

Benabid AL, Benazzouz A, Hoffmann D, Limousin P, Krack P, Pollak P. Long-term electrical inhibition of deep brain targets in movement disorders. *Mov Disord*. 1998;13 Suppl 3:119-25 Clancy, F. Following the leads. Minnesota Medicine 85, April 2002. http://www.mnmed.org/publications/MNMed2002/April/Clancy.html

Phurrough S, Atkinson M, Bridger P, Carino T, Larson W, Schott L. National Coverage Determination Memorandum for Deep Brain Stimulation for Essential Tremor and Parkinson's Disease. http://www.cms.hhs.gov/coverage/download/id21.pdf

Eskandar EN, Flaherty A, Cosgrove GR, Shinobu LA, Barker FG 2nd. Surgery for Parkinson disease in the United States, 1996 to 2000: practice patterns, short-term outcomes, and hospital charges in a nationwide sample. *J Neurosurg*. 2003 Nov; 99(5): 863-71. (abstract saved in BCI as eskandar_dbs-cost.txt)

164 *obsessive compulsive disorder:* Greenberg BD, Rezai AR. Mechanisms and the current state of deep brain stimulation in neuropsychiatry. *CNS Spectr*. 2003 Jul; 8(7): 522-6. Review

164 *she burst out laughing:* Satow T, Usui K, Matsuhashi M, Yamamoto J, Begum T, Shibasaki H, Ikeda A, Mikuni N, Miyamoto S, Hashimoto N. Mirth and laughter arising from human temporal cortex. *J Neurol Neurosurg Psychiatry*. 2003 Jul; 74(7): 1004-5.

165 *Delgado could stimulate:* Delgado, J.M.R. *Physical Control of the Mind.* New York: Harper & Row, 1969.

168 *Memento:* Nolan C (Director). Todd J, Todd S (Producers). 2001. Memento [Motion picture]. Beverley Hills, CA: I Remember Productions.

168 *more than 5 million people:* National Center for Injury Prevention and Control (NCIPC). Traumatic Brain Injury Facts. CDC web site. http://www.cdc.gov/doc.do?id=0900f3ec800081d7

American Stroke Association. About Stroke (web site). http://www.strokeassociation.org/presenter.jhtml?identifier=11402

Thurman DJ, Alverson C, Dunn KA, Guerrero J, Sniezek JE. Traumatic brain injury in the United States: A public health perspective. *J Head Trauma Rehabil*. 1999 Dec; 14(6): 602-15.

170 *like neurons in the hippocampus:* Huang GT. Mind-Machine Merger. *MIT Technology Review* May 2003: 38-45. Available at http://www.technologyreview.com/articles/huang0503.asp

170 *agnosia:* Sacks O. The Man Who Mistook His Wife For a Hat. Simon & Schuster: Summit, NY, 1985.

171 *change the monkey's perceptions:* Huang GT. Mind-machine merger. *MIT Technology Review* May 2003. http://www.technologyreview.com/articles/huang0503.asp

Chapter 10

178 *implants fabricated like microchips:* Wise, KD. Professor Kensall D. Wise. [web site]. http://www.eecs.umich.edu/~wise/ accessed 15 January 2004

178 *16,384 neurons:* Infineon. Neuro-Chip from Infineon Can Read Your Mind - New Findings in Brain Research Expected. [web site] (2003-

02-11)

http://www.infineon.com/cgi/ecrm.dll/jsp/showfrontend.do?lang=EN &content_type=NEWS&content_oid=57550&news_nav_oid=-9979 Accessed 1/21/2004

178 *55 million transistors:* Intel® Pentium® 4 Processor Product Information. [web site] http://www.intel.com/products/desktop/processors/pentium4/ Accessed 1/21/2004

178 *deep brain stimulators:* Eskandar EN, Flaherty A, Cosgrove GR, Shinobu LA, Barker FG 2nd. Surgery for Parkinson disease in the United States, 1996 to 2000: practice patterns, short-term outcomes, and hospital charges in a nationwide sample. *J Neurosurg*. 2003 Nov; 99(5): 863-71

179 *EEG devices:* Wickelgren I. Neuroscience. Power to the paralyzed. *Science*. 2003 Jan 24; 299(5606): 497.

Wickelgren I. Neuroscience. Tapping the mind. *Science*. 2003 Jan 24; 299(5606): 496-9.

180 *limitations of fMRI:* Saini S, Seltzer SE, Bramson RT, Levine LA, Kelly P, Jordan PF, Chiango BF, Thrall JH. Technical cost of radiologic examinations: analysis across imaging modalities. *Radiology*. 2000 Jul; 216(1): 269-72

S.-G. Kim, S.-P. Lee, B. Goodyear, and A. Silva, "Spatial Resolution of BOLD and Other Functional MRI Techniques", in *Medical Radiology - Diagnostic Imaging and Radiation Oncology, Volume Functional MRI* (eds. C. Moonen and P.A. Bandettinni), Springer-Verlag, pp 195-203, 1999.

181 *transcranial magnetic stimulation:* Kloza B. Watching Living Brains (01.21.03). *ScienCentralNEWS* [online serial]. Available at http://www.sciencentral.com/articles/view.php3?language=english&type=&article_id=218391866

Cowey A, Walsh V. Tickling the brain: studying visual sensation, perception and cognition by transcranial magnetic stimulation. *Prog Brain Res*. 2001; 134: 411-25. Review

182 *10 million people:* Lasik and Its Alternatives: an Update. *Med Lett Drugs Ther*. 2004 Jan 19; 46(1174): 5-7

182 *endovascular surgery for the brain:* Boulos AS, Bendok BR, Levy EI, Kim SH, Qureshi AI, Guterman LR, Hopkins LN. Endovascular aneurysm treatment: a proven therapy. *Neurol Res*. 2002; 24 Suppl 1: S71-9. Review.

Bairstow P, Dodgson A, Linto J, Khangure M. Comparison of cost and outcome of endovascular and neurosurgical procedures in the treatment of ruptured intracranial aneurysms. *Australas Radiol.* 2002 Sep; 46(3): 249-51.

183 *125,000 endovascular treatments:* A Patient's Guide to Stroke & Brain Aneurysm, Virginia Mason Hospital Vascular Center

Rosenorn J, Eskesen V, Schmidt K. Unruptured intracranial aneurysms: an assessment of the annual risk of rupture based on epidemiological and clinical data. *Br J Neurosurg*. 1988; 2(3): 369-77. Review

183 *"Our analysis of the brain":* Duke Magazine, January-February 2004, Monkeys Move Matter, Mentally

186 *information technology increases the benefits of human cooperation:* Robert Wright, *Non-Zero*

Chapter 11

193 *"we are snipping the very last weight":* Bill McKibben, Enough

194 *"the human soul yearns for":* Leon Kass, *Life, Liberty, and the Pursuit of Dignity*, page 269

196 *brain evolution:* H.J. Jerison, *Evolution of the Brain and Intelligence*, Academic Press, 1973

References

14248404R00146

Made in the USA
Lexington, KY
21 March 2012